D1807478

Geobotany Studies

Basics, Methods and Case Studies

About the Series

The series includes outstanding monographs and collections of papers on a given topic in the following fields: Phytogeography, Phytosociology, Plant Community Ecology, Biocoenology, Vegetation Science, Eco-informatics, Landscape Ecology, Vegetation Mapping, Plant Conservation Biology and Plant Diversity. Contributions are expected to reflect the latest theoretical and methodological developments or to present new applications at large spatial or temporal scales that could reinforce our understanding of ecological processes acting at the phytocoenosis and vegetation landscape level. Case studies based on large data sets are also considered, provided they support habitat classification refinement, plant diversity conservation or vegetation change prediction. Geobotany Studies: Basics, Methods and Case Studies is the successor to Braun-Blanquetia published by the University of Camerino between 1984 and 2011 with cooperation of Station Internationale de Phytosociologie (Bailleul-France) and Dipartimento di Botanica ed Ecologia (Université de Camerino - Italia) and under the aegis of Societé Amicale Francophone de Phytosociologie, Societé Francaise de Phytosociologie, Rheinold Tuexen Gesellschaft and the Eastern Alpine and Dinaric Society for Vegetation Ecology. This series aims to promote the expansion, evolution and application of the invaluable scientific legacy of the Braun-Blanquetia school.

More information about this series at
http://www.springer.com/series/10526

Part I

Contents

Fabio Conti
Centro Ricerche Floristiche dell'Appennino
Università di Camerino – Parco Nazionale del Gran Sasso e Monti della Laga
Barisciano (AQ)
Italy

Fabrizio Bartolucci
Centro Ricerche Floristiche dell'Appennino
Università di Camerino – Parco Nazionale del Gran Sasso e Monti della Laga
Barisciano (AQ)
Italy

ISSN 2198-2562 ISSN 2198-2570 (electronic)
Geobotany Studies
ISBN 978-3-319-09700-8 ISBN 978-3-319-09701-5 (eBook)
DOI 10.1007/978-3-319-09701-5

Library of Congress Control Number: 2015932230

Springer Cham Heidelberg New York Dordrecht London

Printed on acid-free paper

Springer International Publishing AG Switzerland is part of Springer Science+Business Media (www.springer.com)

Fabio Conti • Fabrizio Bartolucci

The Vascular Flora of the National Park of Abruzzo, Lazio and Molise (Central Italy)

An Annotated Checklist

Introduction

The vascular flora of the Abruzzo, Lazio and Molise National Park has been the subject of more or less extensive research since the first half of the nineteenth century. The first complete list of flora in the protected area, limited to the Abruzzo region, was published in 1959–1960 by Anzalone and Bazzichelli (1960), followed by an update published by Conti (1995) and also including the regions of Lazio and Molise.

The necessity for an updated list of flora incorporating the latest nomenclatural and floristic novelties (especially during the last decade) led to preparation of this flora.

The list of plants was extrapolated from a geographic database including all data from floristic or vegetational references and herbarium specimens concerning the Park area.

This data storage tool was obtained from the database of Abruzzo vascular flora (Conti et al. 2010), adapted to the study area by the addition of the areas of the Park falling in the regions of Lazio and Molise with related floristic and vegetational data. Analysis of the data entered enabled gaps in the floristic knowledge of the Park, such as little or unexplored areas, to be identified, together with the species records requiring confirmation or further study. On the basis of these deductions, field work for the collection of new floristic data was carried out. Verification of the correct identification of herbarium specimens collected in the past, as well as systematic study of critical genera, was also important.

F. Conti, F. Bartolucci, *The Vascular Flora of the National Park of Abruzzo, Lazio and Molise (Central Italy)*, Geobotany Studies, DOI 10.1007/978-3-319-09701-5_1

Geography, Geomorphology and Geology

The National Park falls within three administrative regions (Abruzzo, Lazio and Molise) and covers an area of 104,000 ha, including the external buffer area (Fig. 1). It is a mountainous area in the central part of the Apennines, characterized by a number of chains running in a predominantly North–West, South–East direction, from the mountains overlooking the Fucino to the NW to the Mainarde mountains to the SE. The Park is bounded to the NE by the chain consisting of Montagna Grande, Monte Godi, Serra Rocca Chiarano and Monte Greco and to the SW by the ridge forming the left orographic side of the Liri valley. The highest peaks are Monte Greco (2,285 m), Monte Petroso (2,249 m), Monte Marsicano (2,245 m) and Monte Meta (2,242 m). The rivers flow into the Adriatic and Tyrrhenian seas and involve a number of different basins: the upper Sangro valley (the heart of the Park), the upper valley of the Giovenco river, the Vallelonga valley, the upper valley of the Melfa river (Val Canneto) which then flows into the Liri river and finally the springs and a short stretch of the upper Volturno river.

The geology of the Park is very diversified and consists exclusively of units of the carbonate Lazio-Abruzzo platform and marginally of post-orogenic complex (Vezzani and Ghisetti 1998; D'Andrea et al. 2003; Praturlon et al. 2003). The western sector of the Park between the Liri and Vallelonga valleys is characterized by the Mesozoic internal carbonate platform succession, while in correspondence with the Monte Longana-Serra Lunga complex there are mainly outcrops of whitish limestone with radiolites (upper Cenomanian—Senonian) and dolomitic limestones (limestone dolomite unit, middle Dogger—early Lias). The Vallelonga valley is characterized by alluvial outcrops of sandy-clayey lacustrine deposits and gravelly-sandy fluvial deposits. To the east of the Vallelonga valley, the ridge of Monte Ara dei Merli—Monte Fontecchia is characterized by limestone and dolomitic limestone facies of the platform. The valley of the Giovenco river—upper part of the Sangro river is mainly characterized by clay and sandstone deposits of Val Roveto flysch (Messinian). Eastward meet platform facies evolving to the terms of the margin of the platform as the ridge of Montagna Grande-Monte Marsicano, mainly formed at the base by dolomitic lithologies in stratigraphic succession with

F. Conti, F. Bartolucci, *The Vascular Flora of the National Park of Abruzzo, Lazio and Molise (Central Italy)*, Geobotany Studies, DOI 10.1007/978-3-319-09701-5_2

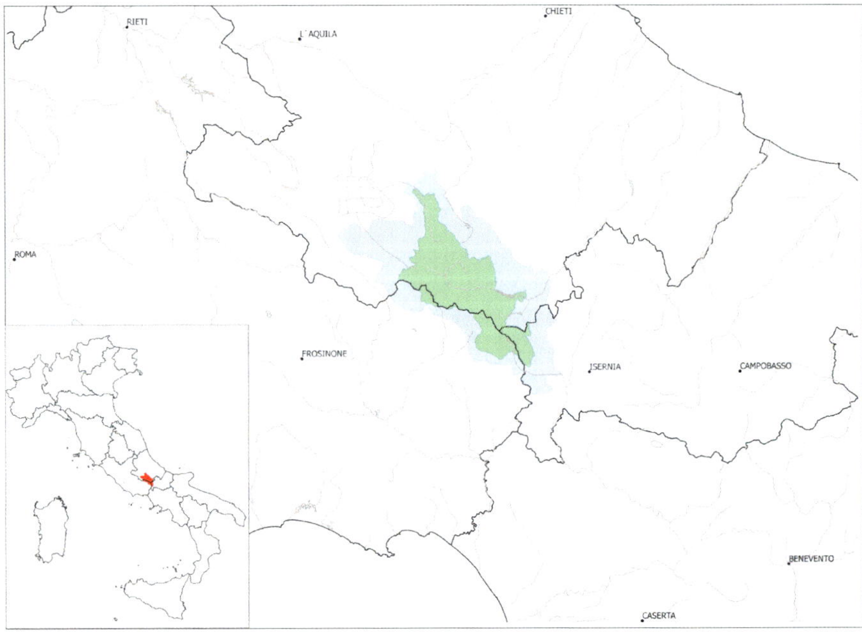

Fig. 1 Map of Abruzzo, Lazio and Molise National Park (green), external buffer area (grey)

palaeodasycladus limestones and in the upper part of the findings from organogenic limestones attributable to the Terratta Formation (Jurassic—Neocomiano). To the east, the Profluo, Tasso and Sagittario valleys are characterized mainly by Anversa degli Abruzzi flysch (Messinian). The Monte Genzana-Monte Greco ridge is characterized by Meso-Cenozoic succession silico-calcareous-marly of basin-slope facies. Finally, in the southern sector of the Park consisting of Monte Meta and the Mainarde mountains there are sequences of carbonate platform and slope facies characterized by calcarenites, detrital limestones and dolomitic limestones.

Climate

Mostly mountainous, the climate parameters of the Park area differ greatly according to altitude. The average annual rainfall ranges from about 650 mm for Anversa degli Abruzzi up to 1,700 mm for some peaks, as shown by the hydrological map (Boni et al. 1986). The maximum values are among the highest in the central Apennines. The data are, however, partial and approximate, lacking meteorological stations at lower altitudes and, from the higher ones, temperature data in particular.

The highest rainfall generally occurs in the autumn with a secondary peak in winter (January) and to a lesser extent in spring (March). The average temperatures are quite low ($\leq$10 °C) and, given the relatively abundant rainfall, do not result in stress from summer aridity, known only for the Barrea station. This stress is, of course, more evident and pronounced in the foothill resort part of the external buffer area with a distinctly Mediterranean climate. No data is, however, available.

F. Conti, F. Bartolucci, *The Vascular Flora of the National Park of Abruzzo, Lazio and Molise (Central Italy)*, Geobotany Studies, DOI 10.1007/978-3-319-09701-5_3

Vegetation Features

The altitude of the National Park of Abruzzo, Lazio and Molise (central Apennines) ranges from 600 to 700 m in the external valleys (Villavallelonga, Alfedena, San Donato Val di Comino, etc.) to the crestline of the mountains, such as Monte Marsicano (2,245 m) and Monte Greco (2,283 m) on the orographic left of the Valle del Sangro and Monte Tranquillo (1,843 m), Monte Capraro (2,100 m) and Monte Meta (2,242 m), on the orographic right.

The Park lies entirely in the eurosiberian region (Conti 2004) with forest vegetation for the most part of deciduous trees of hilly (*Quercus pubescens*, *Q. cerris*, *Ostrya carpinifolia*, *Fraxinus ornus*) and montane zones (*Fagus sylvatica*); the exception is the forests of evergreen sclerophylls (*Quercus ilex*), present in some areas with southern exposure, and those of needle leaves (*Pinus nigra*) of Villetta Barrea, in the Valle del Sangro.

In this altitudinal space, one can recognize the following vegetation belts: hilly (up to 900–1,000 m), montane (from 900–1,000 to 1,800 m), subalpine (from 1,800 to 2,100 m) and alpine (over 2,100 m).

The hilly belt is characterized by deciduous thermophile forests of the *Quercetalia pubescentis* order, with the following associations: flowerish ash (*Melittio melissophylii-Ostryetum carpinfoliae*), scrub oak woods (*Cytiso-Quercetum pubescentis*), turkey oak woods (*Aremonio agrimonioidis-Quercetum cerris* and *Daphno laureolae-Quercetum cerris*).

The flowering ash is limited to the Lazio side of the Park and to a few locations at the entrance of the Vallelonga and the Valle del Sangro, on calcareous substrata; it is always reduced to the state of coppice and interrupted by clearings with secondary meadows of the *Brometalia* order.

The scrub oak woods grow on the slopes of the mountains near the basin of the Fucino, on marly-arenaceous substrata; like the flowering ash woods, they are always reduced to the state of coppice and interrupted by clearings of secondary grazing land of the *Brometalia* order.

F. Conti, F. Bartolucci, *The Vascular Flora of the National Park of Abruzzo, Lazio and Molise (Central Italy)*, Geobotany Studies, DOI 10.1007/978-3-319-09701-5_4

Turkey oak woods are present above all in the middle of the Valle del Sangro, but extensive ones are also found in the high Valle del Sangro, on the right orographic slope.

The landscape of the valley bottom used to be characterized by agriculture fields, but today they are almost completely abandoned, and have been taken over by shrubby vegetation of the *Prunetalia spinosae* order, which forms the classic landscape of closed fields, that is, bordered by hedges or bocage; in some cases the dynamism of the vegetation leads to the formation of woods of trembling poplar of the *Melico uniflorae-Populetum tremulae* association, as at the Gioia Vecchia pass (Pedrotti 1996).

The montane belt is dominated by the great beech forests (Fig. 1), with two associations of the *Fagetalia sylvaticae* order that succeed one another altitudinally, namely *Aremonio agrimonoidis-Fagetum sylvaticae* (900–1,200 m) and *Cardamino kitaibelii-Fagetum sylvaticae* (from 1,200 m to the treeline). The beechwoods for the most part are extensive high forests with continuity throughout the Park, with some large clearings of secondary meadows of the *Brometalia* order (Fig. 2). However, it should be observed that often the belt of beechwoods at the

Fig. 1 Beech forest in Val Cervara (Photo by F. Conti)

Fig. 2 Beech forest with the large clearing of Prato Cardoso (Photo by G. Serafini)

Fig. 3 *Potentilla apennina* and pioneer vegetation in the alpine belt (Photo by F. Conti)

higher altitudes has been eliminated to obtain new grazing areas, today occupied by secondary grazing lands. Most of the beechwoods in the Park are in the dynamic phase of regeneration (Canullo and Pedrotti 1992, 1993).

At the border between the hilly and the montane belts there is a strip of relict pinewoods of black pine (*Pinus nigra*) near Villetta Barrea, on the slopes of the Camosciara.

The subalpine belt, the vegetation of which is very reduced because of grazing, is composed of the Swiss mountain pine wood with the *Polygalo chamaebuxo-Pinetum mugo* and *Orthilio secundae-Pinetum mugo* (Monte Meta) associations, and other shrubby associations spread throughout the mountain massifs of the park, such as: *Helianthemo grandiflori-Juniperetum alpinae* and *Phyteumo orbicularis-Juniperetum alpinae* (Stanisci 1994). The shrubwoods of *Rhamnus alpina* have been attributed to the *Rhamno alpinae-Amelanchieretum ovalis* association, but in part also belong to a new association called *Geranio macrorrhizi-Rhamnetum alpinae* (Conti et Pedrotti, provv.).

The primary meadows of the alpine belt (*Seslerietalia apenninae* order) are spread over the sufficiently high mountains of the park above the treeline and the strip of subalpine shrubs (Bazzichelli and Furnari 1970; Biondi et al. 1992; Di Pietro et al. 2005). The high altitude meadows are important for the Abruzzo chamois (*Rupicapra ornata*), according to the research of Ferrari and Rossi (1985) and Ferrari et al. (1988), which refer in particular to the plant species and types of grazing land preferred by the chamois. In the alpine belt, pioneer vegetation is frequent (Figs. 3, 4, 5 and 6), with a variety of associations such as *Festuco dimorphae-Geranietum macrorrhizi* (Conti and Manzi 1992).

The riverbank vegetation is formed of woods of white willow (*Salicetum albae*) and of black alder (*Aro italici-Alnetum glutinosae*), which are always reduced to a thin, interrupted strip, as along the Sangro River (Pedrotti and Gafta 1996). In the mud along the banks of the artificial lake of Barrea there grows the *Botrydietum granulati* association, formed of algae and liverworts (Aleffi 1992).

Of note for its rarity is the swampy vegetation of *Salicetum apenninae*, present in the Lagozzo, a small lacustrine hollow located in the mountains above Alfedena

Fig. 4 *Mcneillia graminifolia* subsp. *rosanoi* on Mt. Marsicano (Photo by F. Conti)

Fig. 5 Screes on Mt. Marsicano with *Papaver alpinum* subsp. *alpinum* (Photo by F. Conti)

(Spada and Conti 1994; Pedrotti et al. 1996). Another interesting wet environment is Lago Pantaniello, on Monte Greco, which has vegetation of floating and submerged hydrophytes formed of *Potamogeton natans*, *P. lucens* and *P. trichoides* (Naviglio 1984).

In the plain between Opi and Pescasseroli and in a few other locations there are partially flooded wet meadows with the *Hordeo-Ranunculetum velutini* and *Deschampsio-Caricetum distantis* associations (Pedrotti et al. 1992) (Figs. 7 and 8); in the depressed zones of the plain there are also a few strips of *Caricetum elatae*, while in the channels the following associations have been observed: *Nasturtietum officinalis*, *Glycerietum plicatae* and *Gycerio-Sparganietum neglecti*. On the slopes there is also the association of mesophilous meadows of *Cynosuro-Trifolietum repentis* (Fig. 9). The mowing of these meadows is conducted according to very ancient regulations that are still valid today (Manzi 1990).

Finally, of note is the synanthropic vegetation of the areas inhabited by man, composed of many associations including *Conietum maculati*, *Heracleo-Rumicetum obtusifolii*, *Anthriscetum sylvestris*, and *Chaerophylletum aurei*.

Fig. 6 Screes on Mt. Marsicano with *Pedicularis rostratospicata* (Photo by F. Conti)

Fig. 7 Wet meadows with *Euphorbia gasparrini* subsp. *samnitica* in loc. Templo (Photo by F. Conti)

The vegetation of the Park was mapped by Bruno and Bazzichelli (1966, 1968) on a scale of 1:50,000; it includes 18 types of vegetation, corresponding to associations and physiognomic formations; Pignatti (1976) made a map on the scale of 1:25,000 for a sector of the Park that covers Monte Chiarano and surrounding areas.

The Park territory is also included in the maps of all of Italy, such as those by Pedrotti (1991), on a scale of 1:1,000,000 and by Pirone et al. (2010), on a scale of

Fig. 8 Wet meadows in Valle Canale close to Collelongo (Photo by F. Conti)

Fig. 9 Wet meadows and mesophilous meadows with *Cynoglossum apenninum* in loc. Templo (Photo by F. Conti)

1:500,000 and of Europe (Bohn et al. 2000), on a scale of 1:2,500,000; these general maps are useful for understanding on different scales the relationships and contacts between the vegetation of the Park and the surrounding vegetation of the entire central Apennines.

Botanic Exploration

The area of the Abruzzo, Lazio and Molise National Park has been extensively studied, given that the Park has always attracted researchers and promoted new scientific studies. The Park was established in 1923, but the first floristic records date back to the beginning of the 1800s. In a study of the mountains surrounding Scanno, Gravina (1812) was the first to give a list of species present in the Park area. Later Tenore (1831, 1835, 1842) and Tenore and Gussone (1842) published abundant floristic data relating to specific areas such as Monte Meta, the Mainarde mountains, the Chiarano mountains, Monte Greco, Picinisco, Barrea, etc. Some data, mainly for the southern part of the Park such as the Mainarde mountains and Monte Meta, are reported by Terracciano in contributions to the Terra di Lavoro flora (1872, 1873, 1874, 1878, 1890). Falqui (1899) in his "Contribution to the flora of the Liri basin" includes most of Terracciano's reports and covers the area of the Park already studied by Terracciano. An important contribution to the flora of the Park was made by Grande who published a list of plants for the Villavallelonga area (1904) and several floristic notes (Grande 1910, 1913, 1914, 1916, 1924) including many references to the Park area. Also worthy of note are the contributions of Zodda (1931) for the flora of the Mainarde mountains, Vaccari and Wilczek (1940) for the historic sectors of the Park from Gioia dei Marsi to Scanno, the papers of Fiori (1927) and Lusina (1954) and the floristic list deriving from the archived studies of the "Società Botanica Italiana (SBI)" carried out in 1953. These works and the plants collected by Grande provided the basis for the first flora of the Abruzzo National Park by Anzalone and Bazzichelli (1960) listing 1,377 *taxa*. Later, many other botanists studied the Park flora, publishing specific studies, such as on the alpine belt flora (Bazzichelli and Furnari 1970), or individual reports and contributions (Spada 1979; Petriccione 1986, 1988; Conti 1992, 1994; Minutillo 1995). This was followed by publication of a catalogue including 1,912 *taxa* (Conti 1995) and covering a much larger area than that investigated by Anzalone and Bazzichelli (1960), including the successive enlargements of the Park and the external buffer area. The research continued and other findings were published (Orsomando 1975; Scoppola and Modena 1997; Pirone and Tammaro 1997;

F. Conti, F. Bartolucci, *The Vascular Flora of the National Park of Abruzzo, Lazio and Molise (Central Italy)*, Geobotany Studies, DOI 10.1007/978-3-319-09701-5_5

Conti 1998; Conti and Minutillo 1998, 2001; Hennecke and Hennecke 1999; Conti et al. 2002, 2006, 2008, 2011a, b, c; Di Pietro et al. 2004, 2005; Conti and Peruzzi 2006; Peruzzi and Bartolucci 2006; Bartolucci and Peruzzi 2007; Griebl 2010; Conti and Bartolucci 2011a). Also worthy of note is the recent flora of the Sagittario Gorges (Conti and Tinti 2012) which falls partly within the external buffer area of the Park.

Materials and Methods

The following list of flora is based on field surveys carried out from 1999 to 2013 and extensive analysis of relevant literature. The *Herbarium Apenninicum* (APP), Herbarium of the National Park (Pescasseroli) and a few specimens stored in FI, NAP, RO were also consulted to complete the field investigations. The plants collected during our field investigations are stored in the *Herbarium Apenninicum* (APP) at the Apennine Flora Research Center.

The list of flora was extrapolated from the Abruzzo, Lazio and Molise National Park geographic database. The structure of this database was created with File Maker Pro 8.5 software and is relational, with tables relating through identification codes. The tables are:

- the flora of the Abruzzo, Lazio and Molise National Park: giving for each *taxa*, the accepted names, synonyms, endemic, non-native status, conservation status and summary of bibliographic and herbarium data;
- bibliography: reporting for each record in the bibliographic lists, the accepted name of the plant, name used by the author in the publication, synonyms, indicated locality (geo-referenced) and complete bibliographic reference;
- APP Herbarium: reporting for each specimen the accepted name of the plant and complete locality of collection;
- localities: including all the sites mentioned in the records in the bibliographic and APP Herbarium specimens and those derived from the names given in the Istituto Geografico Militare Italiano (IGM) 1:25,000 maps. Each locality is geo-referenced and identified by the corresponding X and Y coordinates (UTM ED50).

The database thus created allows fast comprehensive display and interrogation of a vast amount of data, making them easily manageable for the purpose of proper planning and protection of the area. In addition, the database is designed for integration with a geographic information system (GIS). This allows an area to be defined and a list of *taxa* reported there to be extracted from the database, or

F. Conti, F. Bartolucci, *The Vascular Flora of the National Park of Abruzzo, Lazio and Molise (Central Italy)*, Geobotany Studies, DOI 10.1007/978-3-319-09701-5_6

distribution maps for each species, *taxa* belonging to a specific genus or groups of *taxa* at risk to be generated quickly.

A total of 17,972 records deriving from 252 scientific publications, 9,754 herbarium specimens from the Park area and 2,766 georeferenced localities have been computerized. The field surveys also covered areas bordering on the external buffer area, including the most natural geographical boundaries. Considering this wider area the samples computerized amount to about 12,000.

The systematic order and delimitation of the families follow, for ferns (*Equisetidae*, *Ophioglossidae*, *Polypodiidae*), the classification proposed by Christenhusz et al. (2011a) and for families belonging to the subclasses of the *Magnolidae* and *Pinidae*, Reveal and Chase (2011) and Christenhusz et al. (2011b) respectively. The nomenclature of species and subspecies follows "*An Annotated Checklist of the Italian vascular flora*" (Conti et al. 2005), its integration (Conti et al. 2007) and the update currently underway (Conti et al. in prep.). The status of non-native*taxa* follows the scheme proposed in the recent "Non-Native Flora of Italy" (Celesti-Grapow et al. 2010). We have not listed *taxa* cultivated or introduced for reforestation or ornamental purposes which do not show signs of naturalization.

For each entity, the accepted name, synonyms, preferential environment and frequency of occurrence in the study area (CC = very common; C = common; R = rare; RR = very rare; NC = not confirmed; D = doubtful) are reported.

Others abbreviations or symbols used are as follows:

- New *taxa* for the Park flora: "*"
- Endemic *taxa*: "E"
- Non-native plants: "A" ["NAT" (Naturalized), "INV" (Invasive); "CAS" (Casual)]

Families, genera, species and subspecies are arranged in alphabetical order. *Taxa* of floristic, biogeographical and/or conservation interest are followed by a short note. *Taxa* recorded in literature only and not found during field investigations are indicated in "*italics*".

Part II

Floristic List

A

Adoxaceae

Adoxa moschatellina L. subsp. ***moschatellina*** - *Fagus sylvatica* woods - PC
Sambucus ebulus L. - uncultivated land, ruderal environments - CC
Sambucus nigra L. - humid woods, glades, hedges, ruderal environments - CC
Sambucus racemosa L. - D - Valle di Canneto (Tenore 1835; Tenore and Gussone 1842), Sorgente della Melfa (Lastoria 2000).
Viburnum lantana L. - submontane and montane thermophilous woods and their margins, scrub and hedges, to 1,100 m - C
Viburnum opulus L. - humid woods - R - Scanno (Anzalone and Bazzichelli 1960), Il Lagozzo! (Conti 1994, 1998).
Viburnum tinus L. subsp. ***tinus*** - maquis, woods of thermophilous broadleaves to 800 m - R - Lago di Grotta Campanaro (Spada 1979), Gole del Sagittario! (Conti and Tinti 2012), Monte Falconara!

Alismataceae

*****Alisma lanceolatum*** With. - ditches, sluggish or stagnant waters - R - F. Volturno between "l'Abbazia di Castel S. Vincenzo" and the sources!, Lago Vivo!
Alisma plantago - aquatica L. - ditches, sluggish or stagnant waters - C

Amaranthaceae

A ***Amaranthus albus*** L. - ruderal environments - NAT - Villetta Barrea (Anzalone and Bazzichelli 1960; Viegi et al. 1990 from a specimen collected by Anzalone).

F. Conti, F. Bartolucci, *The Vascular Flora of the National Park of Abruzzo, Lazio and Molise (Central Italy)*, Geobotany Studies, DOI 10.1007/978-3-319-09701-5_7

A ***Amaranthus cruentus*** L. (*A. paniculatus* L.; *A. hybridus* L. subsp. *cruentus* (L.) Thell.) - fields -CAS - S. Biagio (Zodda 1931 sub *A. hybridus* L. var. *patulus* (Bert.)).

A ***Amaranthus deflexus*** L. - ruderal environments - NAT - Mainarde (Zodda 1931; Viegi et al. 1990), Val Fondillo (Viegi et al. 1990).

Amaranthus graecizans L. (*A. angustifolius* Lam.) - T - Medit.(Subcosmop.) - fields - uncultivated land - R - Mainarde (Zodda 1931 sub *A. graecizans* L. subsp. *sylvester* (Desf.))

Amaranthus hybridus L. (*A. chlorostachys* Willd.) - uncultivated land - R - Scanno (Viegi et al. 1990).

A ***Amaranthus hypochondriacus*** L. (*A. hybridus* L. var. *erythrostachys* Moq.) - fields of *Zea mays* - CAS - Vallelonga at Molino (Iamonico 2009).

A ***Amaranthus powellii*** S. Watson subsp. ***powellii*** - ruderal environments - NAT - near Barrea (Iamonico et al. 2011).

A ***Amaranthus retroflexus*** L. - ruderal environments - NAT

Atriplex patula L. - T - Eurasiat. - ruderal environments - PC

Blitum bonus-henricus (L.) C. A. Mey. (*Chenopodium bonus-henricus* L.; *Anserina bonus-henricus* (L.) Dumort.; *Agathophyton bonus-henricus* (L.) Moq.; *Orthosporum bonus-henricus* (L.) T. Nees) - H - Eur. - places where animals gather, montane nitrophilous environments - CC

Chenopodiastrum hybridum (L.) S. Fuentes, Uotila and Borsch (*Chenopodium hybridum* L.) - ruderal environments - C

Chenopodiastrum murale (L.) S. Fuentes, Uotila and Borsch (*Chenopodium murale* L.) - walls and ruderal environments - R - Mainarde (Zodda 1931).

Chenopodium album L. subsp. ***album*** - fields and ruderal environments - CC

Chenopodium ficifolium Sm. - ruderal environments - RR - near Civitella Alfedena! (Conti and Iamonico 2013).

Chenopodium opulifolium Schrad. ex W. D. J. Koch & Ziz - ruderal environments - R - near Barrea! (Conti and Minutillo 1998).

Chenopodium vulvaria L. - ruderal environments - C

Lipandra polysperma (L.) S. Fuentes, Uotila & Borsch (*Chenopodium polyspermum* L.; *Vulvaria polysperma* (L.) Bubani) - fields - C

Oxybasis urbica (L.) S. Fuentes, Uotila & Borsch (*Chenopodium urbicum* L.) - ruderal environments - NC - Settefrati near Sora (Terracciano 1872).

Polycnemum arvense L. - uncultivated land, trampled areas - R - Colle dell'Olmo di Bobbi! (Conti and Minutillo 1998), Villavallelonga between the village and Madonna della Lanna in loc. Cona Rovara!

Amaryllidaceae

*E ***Allium calabrum*** (N. Terracc.) Brullo, Pavone & Salmeri - stony slopes - RR - Scatafosse!

Allium coloratum Spreng. (*A. cirrhosum* Vandelli; *A. carinatum* L. subsp. *pulchellum* Bonnier & Layens) - arid and stony slopes - PC

Allium cupanii Raf. subsp. ***cupanii*** - arid uncultivated land - PC
Allium flavum L. subsp. ***flavum*** - stony meadows - PC
Allium longispathum F. Delaroche (*A. dentiferum* Webb & Berthel.) - arid uncultivated land - R - Cava di Rena!, mouth of Valle Franchitta near Villalago!, Gole del Sagittario, near the old mill! (Conti and Tinti 2012).
Allium lusitanicum Lam. (*A. senescens* L. subsp. *montanum* (Fr.) Holub) - arid and stony montane and subalpine meadows - C
Allium moschatum L. - stony slopes - R - Aia dei Merli (Anzalone and Bazzichelli 1960), Sperone! (Conti 1998; Conti and Minutillo 1998).
Allium nigrum L. (*Caloscordum nigrum* (L.) Banfi & Galasso) - fields - R - near the sources of the river Volturno (Conti 1995).
Allium oleraceum L. subsp. ***oleraceum*** - uncultivated land, humid meadows - PC
Allium pallens L. (*A. coppoleri* Tineo) - arid environments, uncultivated land - NC - Valle di Canneto (Terracciano 1873 sub *A. paniculatum* L.; Falqui 1899 sub *A. pallens* L. b *paniculatum*).
Allium pendulinum Ten. (*Nectaroscordum pendulinum* (Ten.) Galasso & Banfi) - woods - PC
Allium permixtum Guss. (*A. phthioticum* Boiss. & Heldr.; *A. napolitanum* Cirillo var. *breviradium* Halácsy) (Fig. 1) - humid meadows - RR - Valle Chiara!, Passo Godi! (Conti 1995), La Cicerana! (Conti 1998), Monte di Valle Caprara at Vallone Lampazzo!
Allium porrum L. subsp. ***polyanthum*** (Schult. & Schult. f.) Jauzein & J.-M. Tison (*A. polyanthum* Schult. & Schult. f.) - uncultivated land - C
Allium roseum L. (*Nectaroscordum roseum* (L.) Galasso & Banfi) - arid uncultivated land - C
Allium rotundum L. (*A. scorodoprasum* L. subsp. *rotundum* (L.) Stearn; *A. waldsteinii* G. Don) - uncultivated land, ruderal environments - RR - Bisegna!, Cicerana! (Conti 1998), presso Templo in loc. Atessa! (Conti and Minutillo 1998).
Allium saxatile M. Bieb. subsp. ***tergestinum*** (Gand.) Bedalov & Lovric (*A. tergestinum* Gand.) - stony slopes, ridges - PC
Allium schoenoprasum L. (*A. schoenoprasum* L. subsp. *alpinum* (DC.) Čelak.; *A. schoenoprasum* L. subsp. *sibiricum* (L.) K. Richt.) (Fig. 2) - stony slopes in a cirque at around 2,000 m - R - Mt. Argatone!, Serra della Terratta! (Conti 1995, 1998).
Allium sphaerocephalon L. subsp. ***sphaerocephalon*** - arid and stony pastures - CC
Allium tenuiflorum Ten. (*A. pallens* L. subsp. *tenuiflorum* (Ten.) Stearn) - arid meadows, stony glades - C
*****Allium triquetrum*** L. (*Nectaroscordum triquetrum* (L.) Galasso & Banfi) - woods - RR - Valley of Rio Chiaro, near the confluence with the Volturno river!
Allium ursinum L. (*A. ursinum* L. subsp. *ucrainicum* Oksner & Kleopow; *Nectaroscordum ursinum* (L.) Galasso & Banfi) - *Fagus sylvatica* woods - C
Allium vineale L. - uncultivated land, humid meadows - C
Galanthus nivalis L. (*G. imperati* Bertol.) - cool woods - C

Fig. 1 *Allium permixtum* (Photo by F. Bartolucci)

Fig. 2 *Allium schoenoprasum* (Photo by F. Conti)

Narcissus poëticus L. (incl. *Narcissus angustifolius* Curtis ex Haw.; incl. *Narcissus poëticus* subsp. *radiiflorus* (Salisb.) Baker) - montane pastures - C

Narcissus pseudonarcissus L. - uncultivated land - R - near Rocchetta a Volturno! (Conti 1995).

Sternbergia colchiciflora Waldst. & Kit. - arid and stony meadows - R - territory of Villavallelonga, Collelongo (Anzalone and Bazzichelli 1960).

Sternbergia lutea (L.) Ker Gawl. ex Spreng. (*Amaryllis lutea* L.; *Oporanthus luteus* (L.) Herb.) - arid and stony meadows - R - Picinisco (Terracciano 1874), Monte della Rocchetta!, Monte Castelnuovo! (Conti 1992), Fonte Cupa near Castel S. Vincenzo!, Gole del Sagittario! (Conti 1995). Also observed at Pescina! (Conti 1998), outside the Park but close to the buffer external zone.

Anacardiaceae

Cotinus coggygria Scop. - stony slopes, open woods - RR - Vallone del Lacerno! (Conti 1995). Also observed near Casali di Aschi (Conti 1995) outside the Park but close to the buffer external zone. Probably naturalized.
Pistacia terebinthus L. subsp. ***terebinthus*** - arid and stony slopes, maquis - PC

Apiaceae

Aegopodium podagraria L. - cool woods - C
Aethusa cynapium L. subsp. ***elata*** (Friedl. ex Fisch.) Schübl. & G. Martens (*Ae. cynapium* L. subsp. *agrestis* (Wallr.) Dostál) - fields - PC
Ammoides pusilla (Brot.) Breistr. (*A. verticillata* (Duby) Briq.; *Carum ammoides* (L.) Ball; *Petroselinum ammoides* (L.) Rchb.f.; *Ptychotis ammoides* W. D. J. Koch; *Seseli pusilla* Brot.) - scrub, pastures areas and arid uncultivated land - PC
A ***Anethum graveolens*** L. - fields - CAS - Ortona dei Marsi (Bertoloni 1833–1854), Collelongo (Anzalone and Bazzichelli 1960).
Angelica sylvestris L. subsp. ***sylvestris*** - impluviums, humid woods - C
Anthriscus caucalis M. Bieb. (*Caucalis aequicolorum* All.; *Cerefolium anthriscus* (L.) Beck; *Chaerophyllum anthriscus* (L.) Crantz; *Scandix anthriscus* L.; *Torilis anthriscus* (L.) Gaertn.) - ruderal environments, arid uncultivated land - PC
A ***Anthriscus cerefolium*** (L.) Hoffm. (*A. sativa* (Lam.) Besser; *Cerefolium cerefolium* (L.) Britton; *Cerefolium sativum* (Lam.) Besser; *Chaerophyllum cerefolium* (L.) Crantz; *Ch. sativum* Lam.; *Scandix cerefolium* L.; *Selinum cerefolium* (L.) E. H. L. Krause) - ruderal environments - NAT - Pizzone! (Conti 1992), between Pizzone and la Torretta!
Anthriscus nemorosa (M. Bieb.) Spreng. (*Chaerophyllum nemorosum* M. Bieb.) - woods - R - Carrere (Grande 1904 sub *A. sylvestris* Hoffm. var. *nemorosa* Spr.), between Gioia Vecchio and Pescasseroli (Anzalone and Bazzichelli 1960 as *A. silvestris* Hoff *nemorosa* (Spr.)).
Anthriscus nitida (Wahlenb.) Hazsl. - *Fagus sylvatica* woods - PC
Anthriscus sylvestris (L.) Hoffm. subsp. ***sylvestris*** (*Cerefolium sylvestre* (L.) Besser; *Chaerophyllum sylvestre* L.; *Myrrhis sylvestris* (L.) Spreng.) - *Fagus sylvatica* woods, cool meadows - PC

Fig. 3 *Astrantia pauciflora* subsp. *tenorei* (Photo by F. Conti)

Astrantia major L. subsp. ***involucrata*** (W.D.J. Koch) Ces. (*A. major* var. *involucrata* W.D.J. Koch; *A. carinthiaca* Hoppe ex Mert. & W.D.J. Koch; incl.*A. major* subsp. *apenninica* Wörz) - clearings in *Fagus sylvatica* woods - PC

E ***Astrantia pauciflora*** Bertol. subsp. ***tenorei*** (Mariotti) Bechi & Garbari (*Astrantia tenorei* Mariotti) (Fig. 3) - stony pastures at high altitude - C

Athamanta sicula L. - cliffs and walls - R - Pizzone!, Castel S. Vincenzo! (Conti 1992, 1995; Conti et al. 1990).

Berula erecta (Huds.) Coville (*Sium erectum* Huds.) - slow waters - RR - Scanno (Anzalone and Bazzichelli 1960), Zittola! (from a specimen collected by Rosati in Herb. of the Park).

Bifora testiculata (L.) Spreng. (*Coriandrum testiculatum* L.) - fields - NC - Villavallelonga (Grande 1904).

Bunium bulbocastanum L. - stony pastures - C

E *Bunium petraeum* Ten. (*B. alpinum* Waldst. & Kit. subsp. *petraeum* (Ten.) Rouy & E.G. Camus) - D - The records by Conti and Manzi (1992) and Conti (1995) are to be referred to *B. bulbocastanum*. Probably even earlier records for the Park (Tenore 1835 sub *Ligusticum alpinum*, Tenore and Gussone 1842 sub *Ligusticum alpinum* Spr.; Terracciano 1872, 1873; Crugnola 1894 sub *B. alpinum* W. et K.; Anzalone and Bazzichelli 1960; Stanisci 1997) are to be referred to *B. bulbocastanum*.

Bupleurum baldense Turra - arid meadows - C

Bupleurum falcatum L. subsp. ***cernuum*** (Ten.) Arcang. - stony pastures - C

Bupleurum gerardi All. - open woods - RR - between Villetta Barrea and Scanno (Anzalone and Bazzichelli 1960).

Bupleurum praealtum L. (*B. junceum* L., nom. illeg.) - scrub, hedges - PC

E ***Bupleurum rollii*** (Montel.) Moraldo - dry slopes in open woods - R - near Liscia!, Campoli Appennino (Conti 1995), Monte Falconara!

A ***Bupleurum rotundifolium*** L. - fields - NAT - Villavallelonga (Grande 1904), Bisegna (Anzalone and Bazzichelli 1960).

Bupleurum subovatum Link ex Spreng. - fields - R - Picinisco (Tenore and Gussone 1842 sub *Bupleurum protractum*), near the sources of Volturno river! (Conti 1995 sub *Bupleurum lancifolium* Hornem.)

Cachrys ferulacea (L.) Calest. (*Laserpitium ferulaceum* L.; *Prangos ferulacea* (L.) Lindl.) - arid pastures - PC

Carum carvi L. - humid meadows - RR - Piana di Pescasseroli (Pedrotti et al. 1992; Manzi 1993; Manzi and Conti 2002).

Carum heldreichii Boiss. (*C. flexuosum* (Ten.) Nyman, nom. illeg.) - stony slopes - C

Cervaria rivini Gaertn. (incl. *Peucedanum cervaria* (L.) Lapeyr. var. *cervaria*; incl. *P. cervaria* (L.) Lapeyr. var. *microphyllum* Posp.; *Selinum cervaria* L.; *Peucedanum cervaria* (L.) Lapeyr.) - scrub, open woods - R - Difensa (Grande 1904), western slopes of Monte Castelnuovo! (Conti 1995).

Chaerophyllum aureum L. - nitrified environments - C

Chaerophyllum hirsutum L. - *Fagus sylvatica* wood glades, nitrified environments - C

E ***Chaerophyllum magellense*** Ten. - *Fagus sylvatica* wood glades, nitrified environments - C

Chaerophyllum temulum L. (*Ch. temulentum* L.) - glades, ruderal environments - C

Conium maculatum L. subsp. ***maculatum*** - ruderal environments - C

E ***Coristospermum cuneifolium*** (Guss.) Bertol. (*Ligusticum lucidum* Mill. subsp. *cuneifolium* (Guss.) Tammaro) - screes - C

Daucus bicolor Sm. (*D. broteri* Ten.) - NC - Picinisco (Tenore 1835; Tenore and Gussone 1842). The record for Morrone delle Rose (Conti 1995) has to be referred to *D. carota* subsp. *carota*.

Daucus carota L. subsp. ***carota*** (incl. *D. mauritanucus* L.; incl. *D. carota* L. var. *major* Vis.) - uncultivated land, ruderal environments - CC

Daucus carota L. subsp. *drepanensis* (Tod. ex Lojac.) Heywood (*D. gingidium* L.; *D. gingidium* L. subsp. *polygamus* (Gouan) Onno; *D. polygamus* Gouan) - D - Picinisco nella valle del Melfa (Terracciano 1873).

Daucus guttatus Sm. (*D. setulosus* Guss. ex DC.) - arid uncultivated land - RR - Villavallelonga (Tammaro et al. 1988).

Dichoropetalum schottii (Besser ex DC.) Pimenov & Kljuykov (incl. *Peucedanum schottii* Besser var. *petraeum* (Noë) Koch; *P. schottii* Besser ex DC.; *Holandrea*

schottii (Besser ex DC.) Reduron, Charpin & Pimenov) - scrub - RR - Villavallelonga (Grande 1924; Anzalone and Bazzichelli 1960).

Elaeoselinum asclepium (L.) Bertol. subsp. ***asclepium*** - thermophilous stony slopes - PC

Eryngium amethystinum L. - arid pastures - CC

Eryngium campestre L. - arid pastures - C

Falcaria vulgaris Bernh. - fields, uncultivated land - R - Villavallelonga, Case di Maggio (Anzalone and Bazzichelli 1960; Lorito and Veri 1975) and generically recorded for the Park (Sipari 1926 quotes Pirotta; Lusina 1954).

Ferula glauca L. (*Ferula communis* L. subsp. *glauca* (L.) Rouy & E.G. Camus) - arid stony slopes, ruderal environments - R - Castel S. Vincenzo!, Pizzone (Conti 1992).

Ferulago campestris (Besser) Grecescu (*Ferula campestris* Besser; *Ferula ferulago* L.; *Ferulago galbanifera* W. D. J. Koch) - arid pastures - PC

Foeniculum vulgare Mill. subsp. ***piperitum*** (Ucria) Bég. - arid uncultivated land - PC

Foeniculum vulgare Mill. subsp. ***vulgare*** - arid uncultivated land - C

Grafia golaka (Hacq.) Rchb. (*Athamanta golaka* Hacq.; *Hladnikia golaka* (Hacq.) Rchb.f.) - montane meadows - C

Helosciadium nodiflorum (L.) W. D. J. Koch (*Apium nodiflorum* (L.) Lag.; *Sium nodiflorum* L.) - sluggish waters - C

Heracleum orsinii Guss. (*H. sphondylium* L. subsp. *orsinii* (Guss.) H. Neumayer; *H. pyrenaicum* Lam. subsp. *orsinii* (Guss.) Pedrotti & Pignatti) - screes, stony slopes - C

Heracleum sibiricum L. subsp. ***sibiricum*** (*H. sphondylium* L. subsp. *sibiricum* (L.) Simonk.) - open woods - PC

Heracleum sibiricum L. subsp. ***ternatum*** (Velen.) Briq. (*H. ternatum* Velen.; *H. spondylium* subsp. *ternatum* (Velen.) Brummitt) - humid uncultivated land - PC

Heracleum sphondylium L. subsp. *elegans* (Crantz) Schübl. and G. Martens (*H. cordatum* C. Presl; *H. montanum* Schleich. ex Gaudin; *H. pyrenaicum* Lam. subsp. *cordatum* (C. Presl) Pedrotti & Pignatti; *H. protheiforme* Crantz var. *elegans* Crantz) - D - Costa di Cavallaro at Monte Meta (Terracciano 1872; Falqui 1899), Settefrati (Petriglia 2004).

Katapsuxis silaifolia (Jacq.) Reduron, Charpin & Pimenov (*Cnidium apioides* (Lam.) Spreng.; *Laserpitium silaifolium* Jacq.; *Cnidium silaifolium* (Jacq.) Simonk.) - scrub - PC

Laserpitium latifolium L. (incl. *L. latifolium* L. subsp. *asperum* (Crantz) Schübl. & G. Martens) - stony slopes - C

E ***Laserpitium siler*** L. subsp. ***siculum*** (Spreng.) Santang., F. Conti & Gubellini (*L. garganicum* (Ten.) Bertol. subsp. *siculum* (Spreng.) Pignatti; *L. siculum* Spreng.) - stony slopes - C

Myrrhoides nodosa (L.) Cannon (*Chaerophyllum nodosum* (L.) Crantz; *Physocaulis nodosus* (L.) W. D. J. Koch; *Scandix nodosa* L.) - scrub, margins of woods - PC

Oenanthe fistulosa L. - humid meadows - R - Il Pantano (Pedrotti 1983); Montenero Val Cocchiara!

Oenanthe pimpinelloides L. - humid meadows, humid woods - C

Oenanthe silaifolia M. Bieb. - humid meadows - R - Casone del Medico, Sorgenti del Volturno! (Conti 1995), Pantanello (Montenero Val Cocchiara)!

Opopanax chironium (L.) W. D. J. Koch (*Ferulago geniculata* Guss.; *Laserpitium chironium* L.; *Pastinaca opopanax* L.) - arid uncultivated land - C

Oreoselinum nigrum Delarbre (*Athamanta oreoselinum* L.; *Peucedanum oreoselinum* (L.) Moench) - scrub, glades - PC

Orlaya daucorlaya Murb. - NC - generically recorded for the Park (Sipari 1926 sub *Daucus grandiflorus* Scop. var. *daucorlaya* Drude quotes Pirotta; Lusina 1954 sub *Daucus grandiflorus* var. *daucorlaya* Drude).

Orlaya grandiflora (L.) Hoffm. (*Caucalis grandiflora* L.; *Daucus grandiflora* (L.) Scop.) - arid uncultivated land - C

Orlaya platycarpos W. D. J. Koch (*O. kochii* Heywood) - fields - C

Pastinaca sativa L. subsp. ***urens*** (Req. ex Godr.) Čelak. (*P. urens* Req. ex Godr.) - uncultivated land, ruderal environments - C

A ***Petroselinum crispum*** (Mill.) Fuss (*Apium crispum* Mill.; *Carum petroselinum* (L.) Benth.; *P. hortense* Hoffm.; *P. peregrinum* (L.) Lag.; *P. sativum* Hoffm.; *P. vulgare* Lag.) - fields, uncultivated land - CAS

Pimpinella major (L.) Huds. (*P. magna* L.; incl. *P. major* (L.) Huds. subsp. *rubra* (Hoppe) O. Schwarz) - cool woods - C

Pimpinella peregrina L. - grassy uncultivated land - RR - Gole del Sagittario! (Conti and Tinti 2012).

Pimpinella saxifraga L. subsp. ***saxifraga*** (*P. saxifraga* L. subsp. *minor* Weide, nom. illeg.; *P. saxifraga* L. subsp. *montana* Weide) - arid meadows - PC

Pimpinella tragium Vill. (*P. tragium* Vill. subsp. *lithophila* (Schischk.) Tutin) - stony slopes - C

Pleurospermum austriacum (L.) Hoffm. (*Ligusticum austriacum* L.) - NC - Picinisco alla Valle di Canneto (Tenore and Gussone 1842).

Pteroselinum austriacum (Jacq.) Rchb. (*Selinum austriacum* Jacq.; *Peucedanum austriacum* (Jacq.) W. D. J. Koch) - woods - PC

Sanicula europaea L. - woods, mainly *Fagus sylvatica* - CC

Scandix australis L. subsp. ***australis*** - fields, arid meadows - C

Scandix macrorhyncha C. A. Mey. (*S. pecten-veneris* L. subsp. *hispanica* (Boiss.) Bonnier & Layens; *S. pecten-veneris* L. subsp. *macrorhynca* (C. A. Mey.) Rouy & E. G. Camus) - arid pastures - PC

Scandix pecten-veneris L. subsp. ***pecten-veneris*** - fields - C

Seseli libanotis (L.) W. D. J. Koch subsp. ***libanotis*** (*Athamanta libanotis* L.; *Libanotis daucifolia* (Scop.) Rchb.; *Libanotis montana* Crantz) - stony montane pastures - PC

Seseli montanum L. subsp. ***montanum*** - cliffs and stony slopes - C

Seseli peucedanoides (M. Bieb.) Koso-Pol. (*Bunium peucedanoides* M. Bieb.; *Gasparrinia peucedanoides* (M. Bieb.) Bertol.) - montane pastures - RR - Serra Rocca Chiarano (Di Pietro et al. 2004).

E *Seseli polyphyllum* Ten. (*S. montanum* L. subsp. *polyphyllum* (Ten.) P. W. Ball) - NC - Picinisco allo Schioppaturo (Tenore and Gussone 1842).

Seseli tommasinii Rchb. f. (*S. montanum* L. subsp. *tommasinii* (Rchb. f.) Arcang.; *S. viarum* Calest.) - arid meadows, stony slopes - C

Seseli tortuosum L. subsp. ***tortuosum*** - arid uncultivated land - PC

Sison amomum L. - hedges, humid environments - PC

Smyrnium olusatrum L. - humid uncultivated land, ruderal environments - R - presso la chiesa alta di Pizzone! (Conti 1995), Opi! (from a specimen collected by Rosati in the Herb. of the Park).

Tordylium apulum L. - arid meadows, uncultivated land - C

Tordylium maximum L. - arid uncultivated land - PC

Torilis africana Spreng. (*Caucalis purpurea* Ten.; *T. purpurea* (Ten.) Guss.; *T. arvensis* (Huds.) Link subsp. *purpurea* (Ten.) Hayek; *T. africana* Thumb., nom. illeg.; incl. *T. heterophylla* Guss.; incl. *T. africana* Spreng. var. *heterophylla* (Guss.) Reduron) - arid uncultivated land, ruderal environments - PC

Torilis japonica (Houtt.) DC. (*Caucalis japonica* Houtt.) - uncultivated land - C

Torilis leptophylla (L.) Rchb. f. (*Caucalis leptophylla* L.) - uncultivated land - R - Gole del Sagittario! (Conti and Tinti 2012), Gioia Vecchio!

Torilis nodosa (L.) Gaertn. (*Tordylium nodosum* L.) - NC - Mainarde (Zodda 1931).

Trinia dalechampii (Ten.) Janch. (*Meum dalechampii* Ten.) - stony slopes, pastures at high altitude - CC

Trinia glauca (L.) Dumort. subsp. ***glauca*** - cliffs and stony pastures - PC

Turgenia latifolia (L.) Hoffm. (*Caucalis latifolia* L.) - uncultivated land, fields - R - Barrea, Picinisco (Tenore and Gussone 1842), Villavallelonga (Grande 1904), between Collelongo and Trasacco!

Visnaga daucoides Gaertn. (*Daucus visnaga* L.; *Ammi visnaga* (L.) Lam.) - uncultivated land - RR - Valle Venafrana! (Conti 1992).

****Xanthoselinum venetum*** (Spreng.) Soldano & Banfi (*Peucedanum venetum* (Spreng.) W. D. J. Koch) - scrub - RR - Vallone Lampazzo!

Apocynaceae

Vinca major L. subsp. ***major*** - woods, hedges - C

Vinca minor L. - cool woods - PC

Vincetoxicum hirundinaria Medik. subsp. ***hirundinaria*** - pastures, scrub - C

Aquifoliaceae

Ilex aquifolium L. - montane woods, mainly those of *Fagus sylvatica* - C

Araceae

Arisarum proboscideum (L.) Savi (*Arum proboscideum* L.) - humid woods - RR - Valleporcina! (Conti and Minutillo 2001).

Arum cylindraceum Gasp. (*A. lucanum* Cavara & Grande; *A. orientale* M. Bieb. subsp. *lucanum* (Cavara & Grande) Prime) - montane pastures - R - between Gioia dei Marsi and Gioia Vecchio (Vaccari and Wilczek 1940), Pescasseroli (Lusina 1954), Valle di Canneto (Anzalone and Bazzichelli 1960), territory of Villavallelonga (Guarrera and Tammaro 1991), Scanno! (Conti 1998), Frattura (Paolessi and Serafini in verb.).

Arum italicum Mill. subsp. ***italicum*** - open woods, hedges, ruderal environments - CC

Arum maculatum L. - open woods - C

Biarum tenuifolium (L.) Schott subsp. ***tenuifolium*** - arid meadows - R - Pescosolido (Terracciano 1873; Falqui 1899), Collelongo on the Cime di Annamunna (Anzalone and Bazzichelli 1960), Ortona dei Marsi! (Conti 1998; Conti et al. 1999).

Lemna minor L. - stagnant waters - PC

Araliaceae

Hedera helix L. subsp. ***helix*** - woods, hedges - CC

Aristolochiaceae

Aristolochia lutea Desf. (*A. longa* auct. Fl. Ital.) - *Orno-ostryetum* formations, open woods - C

Aristolochia rotunda L. subsp. ***rotunda*** - arid uncultivated land, open woods - R - near the Abbazia of Castel S. Vincenzo! (Conti 1995).

Asarum europaeum L. (*A. europaeum* L. subsp. *caucasicum* (Duch.) Soó; *A. europaeum* L. var. *caucasicum* Duch.; *A. europaeum* L. subsp. *italicum* Kukkonen & Uotila) - cool woods - R - Val Canneto in loc. Acqua Nera (Bazzichelli and Furnari 1970), Lago Montagna Spaccata (Griebl 2010).

Asparagaceae

Anthericum liliago L. - arid pastures - C

Asparagus acutifolius L. - maquis, thermophilous open woods, hedges - C

Asparagus tenuifolius Lam. - woods - RR - Monte Annamunna (Petriccione et al. 1994).

Bellevalia romana (L.) Sweet (*Hyacinthus romanus* L.) - humid meadows - PC

Convallaria majalis L. - open *Fagus sylvatica* woods mostly over 1,500 m - PC

Loncomelos brevistylus (Wolfn.) Dostál (*Ornithogalum pyramidale* auct., non L.) - arid uncultivated land - PC
Loncomelos narbonensis (L.) Raf. (*Ornithogalum narbonense* L.) - uncultivated land - PC
Loncomelos pyrenaicus (L.) Hrouda subsp. ***pyrenaicus*** - mixed woods - PC
*****Muscari botryoides*** (L.) Mill. subsp. ***botryoides*** - meadows - R - Valle del Rio Chiaro, near the confluence with the Volturno river!, loc. Canale near Collelongo (Paris in verb.).
Muscari comosum (L.) Mill. (*Hyacinthus comosus* L.; *Leopoldia comosa* (L.) Parl.) - arid meadows, uncultivated land - C
Muscari neglectum Guss. ex Ten. (*M. atlanticum* Boiss. & Reut.; *M. racemosum* (L.) Mill.; *M. atlanticum* Boiss. & Reut. subsp. *alpinum* (Fiori) Garbari; *M. racemosum* (L.) Mill. var. *alpinum* Fiori) - meadows, pastures - CC
Ornithogalum cfr. *montanum* Cirillo - D - Fosso Macrana (Gioia Vecchio)!
Ornithogalum comosum L. - stony pastures - C
Ornithogalum divergens Boreau - meadows - C - *O. umbellatum* L. shows a high degree of cytological differentiation (Garbari et al. 2007). Only the cytotype with 27 chromosomes seems to be referable to *O. umbellatum* which has never been reported from Abruzzo and accordingly the records of this species are to be referred to *O. divergens.*
*E ***Ornithogalum etruscum*** Parl. subsp. ***etruscum*** - meadows - RR - Rocca Tre Monti!
E ***Ornithogalum exscapum*** Ten. (*O. ambiguum* A. Terracc.; *O. exscapum* Ten. var. *ambiguum* (A. Terracc.) Fiori) - arid pastures - RR - Le Forme!
E ***Ornithogalum orthophyllum*** Ten. subsp. ***orthophyllum*** - arid pastures - PC
Ornithogalum refractum Kit. ex Willd. - arid pastures, arid uncultivated land - PC
Polygonatum multiflorum (L.) All. (*Convallaria multiflora* L.) - *Fagus sylvatica* woods - C
Polygonatum odoratum (Mill.) Druce (*Convallaria odorata* Mill.; *P. officinale* All.) - thermophilous open woods - C
Polygonatum verticillatum (L.) All. (*Convallaria verticillata* L.) - *Fagus sylvatica* woods - PC
Prospero autumnale (L.) Speta (*Scilla autumnalis* L.) - arid and stony meadows - RR - Monte della Rocchetta! (Conti 1992).
Ruscus aculeatus L. - mediterranean and supramediterranean woods - PC
Ruscus hypoglossum L. - cool woods - R - Villavallelonga (Anzalone and Bazzichelli 1960), Colle Stazzo Pavone (Scoppola and Modena 1997).
Scilla bifolia L. - *Fagus sylvatica* woods, pastures - CC

Aspleniaceae

Asplenium adiantum-nigrum L. subsp. *adiantum-nigrum* - woods, shady cliffs - NC - Picinisco allo Schioppaturo (Tenore and Gussone 1842; Fiori 1943).

Asplenium ceterach L. subsp. ***bivalens*** (D. E. Mey.) Greuter & Burdet (*Ceterach officinarum* Willd. subsp. *bivalens* D. E. Mey.; *C. javorkeanum* Vida) - walls and cliffs - CC

Asplenium ceterach L. subsp. ***ceterach*** (*Ceterach officinarum* Willd. subsp. *officinarum*) - cliffs - R - Valle del Giovenco near Pescina! (Conti et al. 2011a). The record for Monte a Mare (Giancola and Stanisci 2006) has probably to be referred to *A. ceterach* subsp. *bivalens*.

Asplenium fissum Kit. ex Willd. - high cliffs - C

Asplenium lepidum C. Presl subsp. ***lepidum*** - cliffs - RR - near Cava di Rena! (Conti and Tinti 2012).

Asplenium onopteris L. - woods, shady cliffs - RR - wood near Casale! (Conti 1992).

Asplenium ruta-muraria L. subsp. *dolomiticum* Lovis & Reichst. (incl. *A. eberlei* D. E. Mey.) - H - S-Eur.-Mont. - D - generically recorded for Marsica (Ferrarini et al. 1986) and according to Marchetti (2004) this taxon is to be consider doubtful in Abruzzo.

Asplenium ruta-muraria L. subsp. ***ruta-muraria*** - cliffs - CC

Asplenium scolopendrium L. subsp. ***scolopendrium*** (*Phyllitis scolopendrium* (L.) Newman subsp. *scolopendrium*) - woods, shady cliffs - C

Asplenium trichomanes L. subsp. ***quadrivalens*** D. E. Mey. - woods, shady cliffs - CC

Asplenium viride Huds. - shady cliffs - C

Asplenium x aprutianum Lovis - shady cliffs - RR - Scanno (Soster 2001).

Asteraceae

E ***Achillea barrelieri*** (Ten.) Sch.Bip. subsp. ***barrelieri*** - stony slopes, firm screes at high altitude - PC - Probably have to be referred here the record of *A. nana* (Sipari 1926 quotes Pirotta; Lusina 1954).

E ***Achillea barrelieri*** (Ten.) Sch.Bip. subsp. ***mucronulata*** (Bertol.) Heimerl (*A. mucronulata* Bertol.; *A. oxyloba* (DC.) Schultz Bip. subsp. *mucronulata* (Bertol.) I. Richardson) - humid cliffs, montane and highmontane stony slopes - R - Montagne di Frattura (Tenore 1831), Gole del Sagittario! (Anzalone 1962; Bazzichelli 1972; Lastoria 2000; Conti and Tinti 2012), Serrone della Terratta (Conti 1995), le Ciminiere - Vallore del Carapale! Monte Argatone! la Plaiuccia!

Achillea collina (Becker ex Wirtg.) Heimerl - pastures - PC

Achillea millefolium L. subsp. ***millefolium*** - pastures - C

Achillea nobilis L. - arid pastures - PC - In Italy have been recorded both the subsp. *nobilis* and the subsp. *neilreichii* (A. Kern.) Velen. (Conti et al. 2005). In Abruzzo only the latter is reported, but we do not indicate any subspecies given the difficult distinction between the two subspecies.

Achillea setacea Waldst. and Kit. - arid pastures - C

E ***Achillea tenorei*** Grande (*Achillea virescens* (Fenzl) Heimerl subsp. *tenorei* (Grande) Bässler) - stony montane and highmontane pastures - C

Adenostyles alpina (L.) Bluff & Fingerh. subsp. ***alpina*** (*A. australis* (Ten.) Nyman; *Cacalia alpina* L. var. *australis* Ten.; *A. alpina* (L.) Bluff & Fingerh. subsp. *australis* (Ten.) Greuter; *A. glabra* (Mill.) DC. subsp. *glabra*) - damp margins of *Fagus sylvatica* woods, in rocky places - C

Anthemis arvensis L. subsp. ***arvensis*** - fields - PC

Anthemis arvensis L. subsp. ***incrassata*** (Loisel.) Nyman (incl. *A. arvensis* L. subsp. *acrochordona* Briq. & Cavill.; *A. clavata* Guss.; *A. gemmelarii* Tineo; *A. incrassata* Loisel.) - uncultivated areas, arid pastures - C

Anthemis cotula L. - uncultivated land, ruins, fields - PC

Anthemis cretica L. subsp. ***columnae*** (Ten.) Franzén (*A. cretica* L. var. *columnae* Ten.) - stony meadows at high altitude - C

Arctium lappa L. - abandoned fields, uncultivated land, ruderal environmnents - PC

Arctium minus (Hill) Bernh. (*A. minus* (Hill) Bernh. subsp. *pubens* (Bab.) Arens) - glades - PC

Arctium nemorosum Lej. - uncultivated land, glades - C

Artemisia absinthium L. - arid uncultivated land, ruderal environments - C

Artemisia alba Turra (incl. *A. alba* Turra subsp. *lobelii* (All.) Hegi) - arid and stony meadows - CC

A ***Artemisia verlotiorum*** Lamotte - humid uncultivated land - NAT

Artemisia vulgaris L. - hedges, ruderal environments - PC

Aster alpinus L. subsp. ***alpinus*** - stony meadows at high altitude - PC

Bellidiastrum michelii Cass. (*Aster bellidiastrum* (L.) Scop.; *Doronicum bellidiastrum* L.) - stony slopes - C

Bellis perennis L. (incl. *B. hybrida* Ten.; *B. pusilla* (N. Terracc.) Pignatti) - uncultivated land, meadows - CC

Bellis sylvestris Cirillo - uncultivated areas, pastures - PC

****Bidens cernuus*** L. - humid meadows - RR - Montenero Val Cocchiara! New species for Molise region.

Bidens tripartitus L. subsp. ***tripartitus*** - riversides, humid meadows - PC

Bombycilaena erecta (L.) Smoljan. (*Micropus erectus* L.) - meadows and arid uncultivated land - PC

Calendula arvensis (Vaill.) L. (*Caltha arvensis* Vaill.; *Calendula micrantha* Tineo & Guss.; *C. bicolor* Raf.; *C. arvensis* L. subsp. *bicolor* (Raf.) Nyman) - arid uncultivated land, fields - C

E ***Carduus affinis*** Guss. subsp. ***affinis*** (Fig. 4) - montane nitrophilous environments - CC

Carduus chrysacanthus Ten. - screes - PC

Carduus collinus Waldst. & Kit. subsp. *collinus* - D - generically recorded for the Park (Sipari 1926 quotes Pirotta; Lusina 1954 sub *C. candicans* var. *collinus* W. et K.).

Carduus defloratus L. subsp. ***carlinifolius*** (Lam.) Ces. (*C. carlinifolius* Lam.) - stony slopes over the wood limit and the *Fagus sylvatica* wood glades - C

Carduus nutans L. subsp. *leiophyllus* (Petrović) Stoj. & Stef. (*C. macrolepis* Peterm.; *C. leiophyllus* Petrović; *C. nutans* L. subsp. *macrolepis* (Peterm.) Kazmi) - D - Settefrati (Val Canneto) (Petriglia 2004).

Fig. 4 *Carduus affinis* subsp. *affinis* (Photo by F. Conti)

Carduus nutans L. subsp. ***nutans*** - uncultivated areas and pastures - C

E ***Carduus nutans*** L. subsp. ***perspinosus*** (Fiori) Arènes (*C. micropterus* (Borbás) Teyber subsp. *perspinosus* (Fiori) Kazmi; *C. nutans* f. *perspinosus* Fiori) - arid pastures - C

Carduus nutans L. subsp. *taygeteus* (Boiss. and Heldr.) Hayek (*C. macrocephalus* Desf. subsp. *inconstrictus* (O. Schwarz) Kazmi; *C. taygeteus* Boiss. and Heldr.) - D - Picinisco (Forestella) (Petriglia 2004).

Carduus pycnocephalus L. subsp. ***pycnocephalus*** - ruderal environments - CC

Carlina acanthifolia All. subsp. ***acanthifolia*** - pastures - C

Carlina acaulis L. subsp. ***caulescens*** (Lam.) Schübl. & G. Martens (*C. acaulis* L. subsp. *simplex* (Waldst. & Kit.) Nyman) - pastures - C

Carlina corymbosa L. - arid pastures - PC

Carlina vulgaris L. subsp. ***spinosa*** (Velen.) Vandas (*C. longifolia* Rchb. var. *spinosa* Velen.) - arid pastures - C

Carpesium cernuum L. - NC - Picinisco allo Schioppaturo (Tenore and Gussone 1842).

Carthamus lanatus L. - uncultivated land, ruderal environments - C

E ***Centaurea ambigua*** Guss. subsp. ***ambigua*** (*C. dissecta* Ten. [non Hill] var. *elata* Ten.) - arid and stony pastures, stony slopes - C

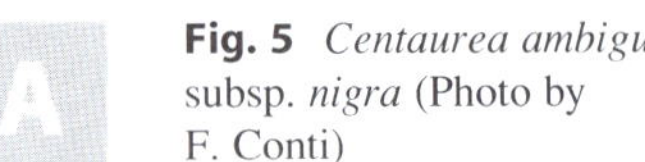

Fig. 5 *Centaurea ambigua* subsp. *nigra* (Photo by F. Conti)

E ***Centaurea ambigua*** Guss. subsp. ***nigra*** (Fiori) Pignatti (*C. delucae* C. Guarino & Rampone; *C. dissecta* Ten. [non Hill] var. *nigra* Fiori) (Fig. 5) - arid and stony pastures, stony slopes at higher altitudes - PC

Centaurea calcitrapa L. (incl. *C. torreana* Ten.) - arid uncultivated land - C

E ***Centaurea ceratophylla*** Ten. subsp. ***ceratophylla*** (*C. rupestris* L. subsp. *ceratophylla* (Ten.) Gugler; *Colymbada ceratophylla* (Ten.) Holub; *Colymbada rupestris* (L.) Holub subsp. *ceratophylla* (Ten.) Banfi, Galasso & Soldano) - cliffs, stony slopes - PC

Centaurea cyanus L. (*Cyanus segetum* Hill) - fields - PC

Centaurea deusta Ten. (incl. *C. alba* L.; incl. *C. alba* L. subsp. *splendens* (L.) Arcang., nom. inval.; incl. *C. deusta* Ten. subsp *concolor* (DC.) Hayek; incl. *C. deusta* Ten. subsp. *divaricata* (Guss.) Matthäs & Pignatti, nom. inval.) - arid meadows - CC

Centaurea jacea L. subsp. ***gaudinii*** (Boiss. and Reut.) Gremli (*C. bracteata* Scop.; *C. amara* L.; *C. gaudinii* Boiss. & Reut.; *Jacea gaudinii* (Boiss. & Reut.) Holub) - arid pastures - CC

Centaurea jacea L. subsp. ***jacea*** - damp meadows - C

Centaurea nigrescens Willd. cfr. subsp. *nigrescens* (*C. carniolica* Host; *C. nigrescens* Willd. subsp. *ramosa* Gugler) - D - Valle Cerasa!, near Liscia, Valle di Mezzo (Conti 1995).

Fig. 6 *Centaurea scannensis* (Photo by F. Conti)

E ***Centaurea nigrescens*** Willd. subsp. ***neapolitana*** (Boiss.) Dostál (*C. neapolitana* Boiss.; *Jacea neapolitana* (Boiss.) Holub; *C. spathulata* Ten.) - meadows, uncultivated land - C

Centaurea scabiosa L. - arid pastures, margins of woods - C - The attribution of the subspecific rank require a revision throughout its range. We started an analysis of material from various origins and first results show some taxa not yet recognized. The name of Linnaeus has not yet been typified but should be based on samples from northern Europe.

E ***Centaurea scannensis*** Anzal., Soldano & F. Conti (Fig. 6) - cliffs, stony slopes - RR - Gole del Sagittario between Anversa and Villalago! (Montelucci 1962; Conti 1995, 1998), Valle Franchitta!

Centaurea solstitialis L. subsp. ***solstitialis*** - ruderal environments - C

Centaurea triumfetti All. (incl. *C. triumfetti* All. subsp. *variegata* (Lam.) Dostal; incl. *Cyanus adscendens* (Bartl.) Sojak; incl. *Cyanus axillaris* J. Presl & C. Presl; incl. *Cyanus canus* auct.; incl. *Centaurea cana* auct.; incl. *Centaurea triumfettii* All. subsp. *nanus* (Ten.) Greuter; incl. *Centaurea axillaris* Willd. var. *nana* Ten.; *Cyanus triumfetti* (All.) Dostál ex Á. Löve & D. Löve.) - arid pastures - CC

Chondrilla chondrilloides (Ard.) H. Karst. (*Lactuca prenanthoides* Scop.; *Ch. prenanthoides* (Scop.) Vill.; *Prenanthes chondrilloides* Ard.; *Willemetia prenanthoides* (Scop.) Gren. & Godr.) - D - between Gioia dei Marsi and Gioia Vecchio (Vaccari and Wilczek 1940 as *Willemetia prenanthoides* Gr. G.). Not confirmed by Anzalone and Bazzichelli (1960) who thought it could be confused with *C. juncea*.

Chondrilla juncea L. - arid uncultivated land and meadows - PC

Cichorium intybus L. (*C. intybus* L. subsp. *glabratum* (C. Presl) Arcang.; incl. *C. intybus* L. subsp. *spicatum* I. Ricci) - arid uncultivated land - C

Cirsium acaulon (L.) Scop. subsp. ***acaulon*** (*Carduus acaulos* L.) - pastures - PC

Cirsium arvense (L.) Scop. - fields, humid uncultivated land - C

Cirsium creticum (Lam.) d'Urv. subsp. ***triumfetti*** (Lacaita) K. Werner - riversides, ditches - PC

E ***Cirsium lobelii*** Ten. - montane pastures - PC

Cirsium oleraceum (L.) Scop. (*Cnicus oleraceus* L.) - humid environments - R - Camosciara!, Villetta Barrea, Civitella Alfedena, Gole di Barrea!, Forca d'Acero (Fiori 1927; Lusina 1954; Anzalone and Bazzichelli 1960; Bruno and Bazzichelli 1966; Conti 1995, 1998; Ballelli et al. 2005).

Cirsium palustre (L.) Scop. - humid environments - R - Camosciara, Val Fondillo, Civitella Alfedena (Lusina 1954; Anzalone and Bazzichelli 1960; Conti 1995, 1998).

E ***Cirsium tenoreanum*** Petr. - montane pastures - C

****Cirsium vulgare*** (Savi) Ten. subsp. ***crinitum*** (DC.) Arènes (*C. crinitum* DC.) - R - Valle di Mezzo and near Collalto! New taxon for Molise region.

Cirsium vulgare (Savi) Ten. subsp. ***vulgare*** - arid uncultivated land, degraded woods - C

Coleostephus myconis (L.) Cass. ex Rchb. f. (*Chrysanthemum myconis* L.; *Myconella myconis* (L.) Sprague; *Myconia myconis* (L.) Briq.; incl. *Coleostephus myconis* (L.) Cass. ex Rchb. f. subsp. *discolor* (Guss.) Arrigoni) - arid and stony pastures - R - surroundings of Gioia Vecchio (1,400–1,600 m), between Pescasseroli and Barrea (Vaccari and Wilczek 1940).

Cota altissima (L.) J. Gay (*Anthemis altissima* L.) - fields, ruins - PC

Cota segetalis (Ten.) Holub (*Anthemis segetalis* Ten.) - fields, uncultivated land - PC

Cota tinctoria (L.) J. Gay subsp. ***australis*** (R. Fern.) Oberpr. and Greuter (*Anthemis tinctoria* L. subsp. ***australis*** R. Fern.) - arid pastures, ruderal environments - C

Cota tinctoria (L.) J. Gay subsp. *tinctoria* (*Anthemis tinctoria* L. subsp. *tinctoria*) - NC - al Monte (Zodda 1931 as *Anthemis tinctoria* L. *typica*).

Cota triumfettii (L.) J. Gay (*Anthemis triumfetti* (L.) DC.; *A. tinctoria* L. var. *triumfettii* L.) - arid slopes, open woods - C

Crepis aurea (L.) Cass. subsp. ***glabrescens*** (Caruel) Arcang. (*C. columnae* (Ten.) Froel.) - meadows at high altitude - C

Crepis biennis L. - meadows - C

Crepis capillaris (L.) Wallr. (*Lapsana capillaris* L.; *Crepis virens* L., nom. illeg.) - uncultivated land, fields - R - San Donato (Falqui 1899), Valle Iannanghera! (Conti 1998; Conti and Minutillo 1998).

Crepis cfr. *corymbosa* Ten. (*Crepis neglecta* L. subsp. *corymbosa* (Ten.) Nyman) - D - Ridge over Castrovalva!

Crepis foetida L. subsp. ***foetida*** - arid uncultivated land - PC

Crepis foetida L. subsp. ***rhoeadifolia*** (M. Bieb.) Čelak. (*C. rhoeadifolia* M. Bieb.) - arid uncultivated land - R - Gole del Sagittario near loc. La Taverna (Conti and Bartolucci 2011a).

Crepis lacera Ten. - arid and stony pastures - C

Crepis neglecta L. - fields, arid uncultivated land - C

Crepis pulchra L. subsp. ***pulchra*** - ruderal environments - C

Crepis pygmaea L. (Fig. 7) - screes - PC

A ***Crepis sancta*** (L.) Babc. subsp. ***nemausensis*** (P. Fourn.) Babc. (*C. bifida* (Vis.) Muschl.; *C. sancta* auct. Fl. Ital.; *Lagoseris nemausensis* (Gouan) W. D. J. Koch;

Fig. 7 *Crepis pygmaea* (Photo by F. Conti)

L. sancta (L.) K. Malý; *Pterotheca nemausensis* (Gouan) C. A. Mey.; *Pterotheca sancta* (L.) K. Koch) - arid uncultivated land, ruderal environments - NAT

Crepis setosa Haller f. - fields, uncultivated land - PC

Crepis vesicaria L. subsp. ***vesicaria*** - arid uncultivated land - CC

Crupina crupinastrum (Moris) Vis. - arid meadows - PC

Crupina vulgaris Cass. - arid meadows - PC

Dittrichia graveolens (L.) Greuter (*Cupularia graveolens* (L.) Gren. & Godr.; *Erigeron graveolens* L.; *Inula graveolens* (L.) Desf.) - abandoned fields, arid uncultivated land - R - Mainarde (Zodda 1931).

Dittrichia viscosa (L.) Greuter subsp. ***viscosa*** (*Inula viscosa* (L.) Aiton subsp. *viscosa*) - ruderal environments, arid uncultivated land - PC

Doronicum columnae Ten. (*Doronicum cordatum* auct. Fl. Ital.) - woods, shady stony slopes - CC

Doronicum pardalianches L. - D - Settefrati (Terracciano 1878; Falqui 1899).

Echinops ritro L. subsp. ***ritro*** - arid pastures, stony slopes - R - Monti di Chiarano (Tenore 1835), Villavallelonga (Anzalone and Bazzichelli 1960), Fossato di Rosa (Petriccione et al. 1994).

Echinops sphaerocephalus L. subsp. ***sphaerocephalus*** - arid pastures and uncultivated areas - C

Erigeron acris L. subsp. ***acris*** - pastures, arid uncultivated areas - PC

Erigeron acris L. subsp. ***angulosus*** (Gaudin) Vacc. (incl. *E. acris* L. subsp. *droebachensis* (O. F. Müll.) Arcang.; *E. angulosus* Gaudin) - meadows at high altitude - PC

Erigeron alpinus L. (incl. *E. alpinus* L. subsp. *intermedius* (Rchb.) Pawll.) - meadows at high altitude - C

Erigeron atticus Vill. (Fig. 8) - grassy - stony slopes - R - Valle dell'Altare (Conti 1995), foot of Mt. Petroso, Valle Cupella, Lago Vivo (Conti 1998), Mt. Cavallo (Conti and Minutillo 1998).

A ***Erigeron bonariensis*** L. (*Conyza bonariensis* (L.) Cronq.) - arid uncultivated land - NAT - Gole del Sagittario! (Conti and Tinti 2012).

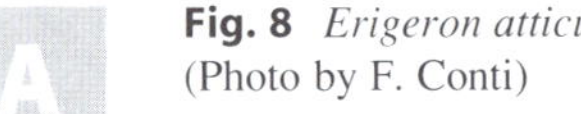

Fig. 8 *Erigeron atticus* (Photo by F. Conti)

A ***Erigeron canadensis*** L. (*Conyza canadensis* (L.) Cronq.) - naturalized, ruderal environments - NAT

Erigeron epiroticus (Vierh.) Halácsy (*Trimorpha epirotica* Vierh.) - meadows at high altitude - C

Erigeron glabratus Bluff & Fingerh. (*E. polymorphus* Scop.) - meadows at high altitude - PC

Erigeron uniflorus L. - meadows at high altitude - PC

Eupatorium cannabinum L. subsp. ***cannabinum*** - watercourses, humid environments - CC

Filago eriocephala Guss. - arid uncultivated land - R - generically recorded for the Park (Anzalone and Bazzichelli 1960).

Filago germanica (L.) Huds. (*Gnaphalium germanicum* L.; *F. vulgaris* Lam.) - NC - Picinisco (Tenore and Gussone 1842).

Filago pyramidata L. - arid meadows, ruderal environments - R - near S. Biagio (Zodda 1931), near the Abbazia of Castel S. Vincenzo! (Conti 1995), Monte S. Angelo!

Galactites tomentosus Moench (*G. pumila* Porta; *G. elegans* (All.) Soldano; *Lupsia galactites* (L.) Kuntze) - arid uncultivated land, ruderal environments - R - Cocullo (Greco and Petriccione 1989).

A ***Galinsoga parviflora*** Cav. - ruderal environments - NAT - Barrea! (Conti 1998; Conti and Minutillo 1998).

Geropogon hybridus (L.) Sch.Bip. (*Tragopogon hybridus* L.; *Geropogon glaber* L.) - uncultivated areas, arid pastures - R - near Ponte Nuovo (Scapoli) (Hennecke and Hennecke 1999)

Glebionis segetum (L.) Fourr. (*Chrysanthemum segetum* L.) - fields - R - Monte Malpasso (Petriccione et al. 1994).

Gnaphalium diminutum Braun-Blanq. (*G. hoppeanum* W. D. J. Koch subsp. *magellense* (Fiori) Strid; *G. supinum* L. f. *magellense* Fiori) - snowbed meadows - C

Gnaphalium sylvaticum L. (*Omalotheca sylvatica* (L.) Sch.Bip. & F. W. Schultz) - montane pastures - C

Gnaphalium uliginosum L. (*Filaginella uliginosa* (L.) Opiz) - riversides and ponds - RR - Serra Rocca Chiarano (Di Pietro et al. 2004).

Hedypnois rhagadioloides (L.) F. W. Schmidt (*Hyoseris cretica* L.; *Hedypnois cretica* (L.) Dum. Cours.) - arid uncultivated areas - RR - near Campoli Appennino (Minutillo, in litt.).

A ***Helianthus annuus*** L. - uncultivated areas - CAS

A ***Helianthus tuberosus*** L. - uncultivated areas - NAT

Helichrysum italicum (Roth) G. Don subsp. ***italicum*** - arid meadows, garigue - CC

Helminthotheca echioides (L.) Holub (*Picris echioides* L.) - fields, arid pastures - C

Hieracium acanthodontoides Arv.-Touv. & Belli - stony slopes partially shaded - PC

Hieracium amplexicaule L. subsp. ***berardianum*** (Arv.-Touv.) Zahn - cliffs - C

Hieracium bifidum Kit. ex Hornem. subsp. ***caesiiflorum*** (Almq. ex Norrl.) Zahn - open *Fagus sylvatica* woods, stony slopes - PC

Hieracium bifidum Kit. ex Hornem. subsp. ***nummulariifolium*** Gottschl. - open *Fagus sylvatica* woods, stony slopes - C

Hieracium bifidum Kit. ex Hornem. subsp. ***stenolepis*** (Lindeb.) Zahn - open *Fagus sylvatica* woods, stony slopes - C

Hieracium chlorifolium Arv.-Touv. subsp. ***rendinaricum*** Gottschl. - stony slopes - R - Monte San Nicola (Gottschlich 2009), Pietre Rosse!

Hieracium glabratum Hoppe ex Willd. - stony slopes - RR - Monte Forcellone! (Conti 1995).

Hieracium glaucum All. (incl. *H. willdenowii* (Monn.) Griseb.) - NC - Picinisco (Terracciano 1873, 1890; Falqui 1899).

Hieracium huetianum Arv.-Touv. - stony slopes - PC

Hieracium humile Jacq. subsp. ***brachycaule*** Zahn - cliffs - C

Hieracium hypochoeroides Gibson subsp. ***bifidopsis*** (Zahn) Greuter - stony slopes - RR - Passo Godi (Gottschlich 2009).

Hieracium hypochoeroides Gibson subsp. ***lithophilum*** (Arv.-Touv.) Greuter - stony slopes - RR - Passo Godi (Gottschlich 2009).

Hieracium hypochoeroides Gibson subsp. ***pallidopsis*** Gottschl. - stony slopes - RR - Passo Godi (Gottschlich 2009).

Hieracium hypochoeroides Gibson subsp. ***potamogetifolium*** Gottschl. - stony slopes - PC

Hieracium jurassicum Griseb. subsp. ***subperfoliatum*** (Arv.-Touv.) Greuter - *Fagus sylvatica* woods - PC

E ***Hieracium latilepidotum*** Gottschl. - stony slopes - RR - Monte Greco - Monte Chiarano! (Gottschlich 2009).

E ***Hieracium marsorum*** Gottschl. - cliffs - RR - Gioia Vecchio! (Gottschlich 2009).

Hieracium murorum L. subsp. ***pleiotrichum*** (Zahn) Zahn - open *Fagus sylvatica* woods - CC

Hieracium naegelianum Pančić subsp. ***andreae*** (Degen & Zahn) Zahn - stony slopes - PC

Hieracium pellitum Fr. (incl. *H. pseudolanatum* Arv.-Touv.; incl. *H. jordanii* Arv.-Touv.) - stony slopes - PC

Hieracium pilosum Schleich. ex Froel. subsp. ***portae*** (Nägeli & Peter) Gottschl. - stony slopes - C

Hieracium prenanthoides Vill. subsp. ***perfoliatum*** (Froel). Fr. - *Fagus sylvatica* woods - RR - between Villetta Barrea and Passo Godi (Gottschlich 2009).

Hieracium prenanthoides Vill. subsp. ***stupposifolium*** Gottschl. - *Fagus sylvatica* woods - C

E ***Hieracium profetanum*** Belli - open woods - RR - Gole del Sagittario! (Conti and Tinti 2012).

Hieracium racemosum Waldst. & Kit. ex Willd. subsp. ***alismatifolium*** (Posp.) Zahn - open woods - R - between Villetta Barrea and Passo Godi, Alfedena (Gottschlich 2009).

Hieracium racemosum Waldst. & Kit. ex Willd. subsp. ***barbatum*** (Froel.) Zahn - open woods - R - between Villetta Barrea and Passo Godi, Alfedena (Gottschlich 2009).

Hieracium racemosum Waldst. & Kit. ex Willd. subsp. ***caramanicum*** (Zahn) Zahn - open woods - C

Hieracium racemosum Waldst. & Kit. ex Willd. subsp. ***pulmonariifolium*** Gottschl. - open woods - C

Hieracium sabaudum L. (incl. *H. sabaudum* L. subsp. *sublactucaceum* Zahn) - woods - R - Valleporcina! (Conti 1995), presso S. Giuseppe tra S. Biagio Saracinisco ed Atina (Terracciano 1872).

Hieracium schmidtii Tausch subsp. ***brunelliforme*** (Arv.-Touv.) O. Bolòs and Vigo - stony slopes - PC

Hieracium schmidtii Tausch subsp. ***crinitisquamum*** Gottschl. (incl. *H. pallidum* Biv. - Bern.f.; incl. *H. schmidtii* Tausch subsp. *brunelliforme* (Arv. - Touv.) O. Bolòs & Vigo) - stony slopes - PC

Hieracium schmidtii Tausch subsp. ***subcomatulum*** (Zahn) O. Bolòs and Vigo - stony slopes - RR - Vallone Lampazzo (Gottschlich 2009).

Hieracium scorzonerifolium Vill. subsp. ***flexuosum*** Nägeli & Peter - stony slopes - RR - La Meta (Gottschlich 2009).

Hieracium scorzonerifolium Vill. subsp. ***scorzonerifolium*** - stony slopes - RR - Forca Resuni (Gottschlich 2009).

Hieracium villosum Jacq. subsp. ***villosum*** - stony slopes - PC

Hyoseris radiata L. (*Thlipsocarpus boeticus* Kunze) - NC - dintorni di Gioia Vecchio (Vaccari and Wilczek 1940).

Hyoseris scabra L. - arid meadows - RR - between Lagone and Mastrogiovanni! (Conti and Minutillo 2001).

Hypochaeris cretensis (L.) Bory & Chaub. - arid pastures - C

Hypochaeris laevigata (L.) Ces., Pass. & Gibelli (*H. laevigata* (L.) Ces., Pass. & Gibelli) - NC - Mainarde (Zodda 1931).

Hypochaeris radicata L. (incl. *H. radicata* L. subsp. *platylepis* (Boiss.) Jahand. & Maire) - NC - Picinisco (Tenore and Gussone 1842).

E ***Hypochaeris robertia*** (Sch.Bip.) Fiori (*Achyrophorus robertia* Sch.Bip.; *Robertia taraxacoides* (Loisel.) DC.; *Seriola taraxacoides* Loisel.) - screes, stony slopes - C

Inula conyzae (Griess.) DC. (*Aster conyzae* Griess.) - open woods - C

Inula helenium L. - humid meadows, ditches - R - between Pizzone and loc. La Cartiera!, Sorgenti del Volturno! (Conti 1995).

Inula hirta L. - arid meadows - R - Monte S. Marcello (Falqui 1899), Villavallelonga (Anzalone and Bazzichelli 1960), Lago Montagna Spaccata, Scontrone (Griebl 2010).

Inula montana L. - arid pastures - C

Inula salicina L. (incl. *I. aspera* Poir.) - humid meadows - C

E ***Jacobaea alpina*** (L.) Moench subsp. ***samnitum*** (Nyman) Peruzzi (*Senecio samnitum* (Nyman) Greuter; *Cineraria cordifolia* Gouan var. *samnitum* Nyman; *Jacobaea samnitum* (Nyman) B. Nord. & Greuter) - humid small valleys in pastures or in clear areas of *Fagus sylvatica* woods - C

Jacobaea erratica (Bertol.) Fourr. (*Senecio aquaticus* Hill var. *barbareifolius* (Krock.) Wimm. & Grab.; *S. erraticus* Bertol.; *S. erraticus* Bertol. subsp. *barbareifolius* (Wimm. & Grab.) Beger) - humid and shady environments - PC

Jacobaea erucifolia (L.) G. Gaertn., B. Mey. & Scherb. subsp. ***erucifolia*** (*Senecio erucifolius* L. subsp. *erucifolius*) - humid uncultivated areas - C

Jurinea mollis (L.) Rchb. subsp. ***mollis*** (incl. *J. mollis* subsp. *mollis* f. *erectobracteata* F. Conti) - stony pastures, stony slopes - PC

Klasea nudicaulis (L.) Fourr. (*Serratula nudicaulis* (L.) DC.) - montane stony pastures - PC

Lactuca muralis (L.) Gaertn. (*Cicerbita muralis* (L.) Wallr.; *Mycelis muralis* (L.) Dumort.; *Prenanthes muralis* L.) - cool glades - C

Lactuca perennis L. - stony slopes - PC

Lactuca saligna L. - uncultivated land - PC

Lactuca serriola L. - uncultivated land - PC

Lactuca viminea (L.) J. Presl & C. Presl subsp. ***chondrilliflora*** (Boreau) St.-Lag. (*L. chondrilliflora* Boreau) - arid and stony slopes - C

Lactuca viminea (L.) J. Presl & C. Presl subsp. *viminea* - NC - al Monte (Zodda 1931 sub *Lactuca viminea* (L.) J. et C. Presl *typica*).

A

Lactuca virosa L. - uncultivated land - C
Lapsana communis L. subsp. ***communis*** - open woods, fields - C
Leontodon crispus Vill. subsp. ***crispus*** (*L. crispus* Vill. subsp. *asper* auct. Fl. Ital.) - stony pastures - C
Leontodon hispidus L. subsp. ***hispidus*** - meadows - C
E ***Leontodon intermedius*** Huter, Porta & Rigo - arid and stony meadows - PC
Leontodon rosanii (Ten.) DC. (*L. villarsii* auct. Fl. Ital.; *Apargia rosanii* Ten.; *L. hirtus* auct. Fl. Ital.) - arid pastures - PC
Leontopodium nivale (Ten.) Huet ex Hand.-Mazz. subsp. ***nivale*** (*L. alpinum* Cass. subsp. *nivale* (Ten.) Tutin) - stony mountain tops - RR - Serra Rocca Chiarano (Petriccione 1986; Lastoria 2000).
Leucanthemopsis alpina (L.) Heywood (*Chrysanthemum alpinum* L.; *Chrysanthemum minimum* Vill.; incl. *L. alpina* (L.) Heywood subsp. *minima* (Vill.) Holub; incl. *L. minima* (Vill.) Marchi; *Pyrethrum alpinum* (L.) Schrank; *Tanacetum alpinum* (L.) Sch.Bip.) - D - Monte Meta (Terracciano 1872; Falqui 1899), Picinisco a Forca dei Fiori presso lo Zaffineto (Terracciano 1873). Most probably to be excluded.
Leucanthemum adustum (W. D. J. Koch) Gremli (*Chrysanthemum adustum* W. D. J. Koch) - montane pastures - R - Monte Genzana!, Picinisco in Valle di Canneto (Tenore and Gussone 1842).
E ***Leucanthemum coronopifolium*** Vill. subsp. ***tenuifolium*** (Guss.) Vogt & Greuter (*L. ceratophylloides* (All.) Nyman subsp. *tenuifolium* (Guss.) Bazzich. & Marchi; *L. tenuifolium* (Guss.) Gamisans; *Pyrethrum tenuisectum* Bertol.) - montane stony slopes at high altitude - PC
Leucanthemum heterophyllum (Willd.) DC. (*Chrysanthemum heterophyllum* Willd.) - montane pastures - RR - Monte della Corte (Gruppo del Monte Marsicano) (Stanisci 1997).
Leucanthemum ircutianum Turcz. ex DC. subsp. ***asperulum*** (N. Terracc.) Vogt & Greuter (*L. vulgare* Lam. var. *asperulum* N. Terracc.) - montane pastures - RR - Fontitune, Picinisco (Marchi 1972; Marchi and Illuminati 1974).
Leucanthemum pallens (Perreym.) DC. (*Chrysanthemum pallens* Perreym.) - fields, meadows, uncultivated land - PC
E ***Leucanthemum tridactylites*** (A. Kern. & Huter) Huter, Porta & Rigo (*L. atratum* (Jacq.) DC. subsp. *tridactylites* (A. Kern. & Huter) Heywood; *Tanacetum tridactylites* A. Kern. & Huter; *Chrysanthemum tridactylites* Fiori) (Fig. 9) - stony slopes - C
Leucanthemum vulgare (Vaill.) Lam. subsp. ***vulgare*** (*L. praecox* Horvatic) - meadows, glades - PC
Matricaria chamomilla L. (*Chamomilla recutita* (L.) Rauschert; *Matricaria recutita* L.) - fields, ruderal environments - PC
A ***Matricaria discoidea*** DC. - ruderal environments, places where animals gather - NAT
Onopordum acanthium L. subsp. ***acanthium*** - uncultivated land - PC
Onopordum illyricum L. subsp. ***illyricum*** - uncultivated land - PC

Fig. 9 *Leucanthemum tridactylites* (Photo by F. Conti)

Pallenis spinosa (L.) Cass. subsp. ***spinosa*** (*Asteriscus spinosus* (L.) Sch.Bip.; *Buphthalmum spinosum* L.) - arid meadows - RR - presso Liscia! (Conti 1995).

Petasites albus (L.) Gaertn. (*Tussilago alba* L.) - NC - Picinisco at the Cartiera (Tenore and Gussone 1842).

Petasites hybridus (L.) G. Gaertn., B. Mey. & Scherb. subsp. ***hybridus*** (*P. officinalis* Moench; *Tussilago hybrida* L.) - watercourses - C

Picnomon acarna (L.) Cass. (*Cirsium acarna* (L.) Moench; *Cnicus acarna* (L.) L.) - arid uncultivated land - R - Villavallelonga (Grande 1904), Monte Genzana!, Frattura!

Picris hieracioides L. subsp. ***hieracioides*** (*P. spinulosa* Bertol. ex Guss.; *P. hieracioides* subsp. *spinulosa* (Bertol. ex Guss.) Arcang.; P. setulosa Guss.; *P. hieracioides* subsp. *setulosa* (Guss.) Arcang.) - arid uncultivated land - C - The nomenclature according to Slovák et al. (2012).

Picris hieracioides L. subsp. ***umbellata*** (Schrank) Ces. (*P. grandiflora* Ten.; *P. hieracioides* subsp. *grandiflora* (Ten.) Arcang) - stony slopes, upper margins of Fagus sylvatica woods - PC - The nomenclature according to Slovák et al. (2012).

*E ***Pilosella calabra*** (Nägeli & Peter) Soják (*Hieracium calabrum* Nägeli & Peter) - stony slopes - RR - Monte Mattone, near Pizzone!

*E ***Pilosella corvigena*** (Gottschl.) Gottschl. (*Hieracium corvigenum* Gottschl.) - stony slopes - RR - between Bisegna and S. Sebastiano (Gottschlich 2009).

Pilosella cymosa (L.) F. W. Schultz & Sch.Bip. subsp. ***sabina*** (Sebast.) H. P. Fuchs (*Hieracium cymosum* L. subsp. *sabinum* (Seb.) Nägeli & Peter) - montane and subalpine pastures - C

Pilosella hoppeana (Schult.) F. W. Schultz & Sch.Bip. (*Hieracium macranthum* (Ten.) Ten.; *H. pilosella* L. var. *macranthum* Ten.; *H. hoppeanum* Schult.) - pastures - PC

Pilosella lactucella (Wallr.) P. D.Sell & C. West subsp. ***nana*** (Scheele) M. Laínz (*Hieracium lactucella* Wallr. subsp. *nanum* (Scheele) P. D.Sell) - mainly pastures with *Nardus stricta* - PC

Pilosella officinarum Vaill. (*Hieracium pilosella* L.) - arid pastures - CC

Pilosella piloselloides (Vill.) Soják (*Hieracium florentinum* All.; *H. piloselloides* Vill.) - arid meadows - C

Podospermum canum C. A. Mey. (*Scorzonera jacquiniana* (W. D. J. Koch) Boiss.; *Podospermum jacquinianum* W. D. J. Koch; *Scorzonera cana* (C. A. Mey.) O. Hoffm.) - arid pastures - C

Podospermum laciniatum (L.) DC. subsp. *decumbens* (Guss.) Gemeinholzer & Greuter (*P. resedifolium* (L.) DC.; *Scorzonera resedifolia* Retz.; *S. laciniata* L. subsp. *decumbens* (Guss.) Greuter; *S. calcitrapifolia* Vahl var. *decumbens* Guss.; *S.calcitrapifolia* Vahl; *S. intermedia* Guss.; *Podospermum calcitrapifolium* (Vahl) DC.) - D - surroundings of Gioia Vecchio (Vaccari and Wilczek 1940).

Podospermum laciniatum (L.) DC. subsp. ***laciniatum*** (*Scorzonera laciniata* L.) - pastures - PC

Prenanthes purpurea L. - *Fagus sylvatica* woods - C

Ptilostemon strictus (Ten.) Greuter (*Cirsium strictum* (Ten.) Link; *Cnicus strictus* Ten.) - open woods and margins - C

Pulicaria dysenterica (L.) Bernh. (*Inula dysenterica* L.) - abandoned fields, ruderal environments - C

Reichardia picroides (L.) Roth (*Picridium vulgare* Desf.; *Scorzonera picroides* L.) - thermophilous stony slopes, arid uncultivated land, walls - PC

Rhagadiolus stellatus (L.) Gaertn. (*R. edulis* Gaertn.) - uncultivated areas, arid pastures, fields, hedges - PC

E *Rhaponticoides centaurium* (L.) M. V. Agab. & Greuter (*Centaurea centaurium* L.) - D - generically recorded for the Park (Rovesti and Rovesti 1934).

Scolymus hispanicus L. subsp. ***hispanicus*** - arid uncultivated land - RR - Gole del Sagittario! (Conti and Tinti 2012).

Scorzonera austriaca Willd. - stony pastures - RR - Monte Rosa Pinnola! (Conti 1995, 1998).

Scorzoneroides autumnalis (L.) Moench (incl. *Leontodon autumnalis* L. subsp. *pratensis* (Link) Arcang.; *L. autumnalis* L.; incl. *Scorzoneroides autumnalis* (L.) Moench subsp. *borealis* (Ball) Greuter) - humid meadows - PC

Scorzoneroides cichoriacea (Ten.) Greuter (*Apargia cichoriacea* Ten.; *Leontodon cichoriaceus* (Ten.) Sanguin.) - arid pastures, glades - C

Fig. 10 *Scorzoneroides montana* subsp. *breviscapa* (Photo by F. Conti)

E ***Scorzoneroides montana*** (Lam.) Holub subsp. ***breviscapa*** (DC.) Greuter (*Leontodon montanus* Lam. subsp. *breviscapus* (DC.) Cavara and Grande; *L. croceus* Haenke var. *breviscapus* DC.) (Fig. 10) - screes - PC

Senecio doronicum (L.) L. - stony pastures - C

A ***Senecio inaequidens*** DC. - arid uncultivated land, ruderal environments, rarely in pastures - INV

E ***Senecio ovatus*** (G. Gaertn., B. Mey. & Scherb.) Willd. subsp. ***stabianus*** (Lacaita) Greuter (*S. nemorensis* L. subsp. *stabianus* (Lacaita) Pignatti; *S. stabianus* Lacaita) - glades in *Fagus sylvatica* woods - PC

E ***Senecio scopolii*** Hoppe & Hornsch. subsp. ***floccosus*** (Bertol.) Greuter (*S. tenorei* Pignatti) - stony pastures - C

Senecio squalidus L. subsp. ***rupestris*** (Waldst. & Kit.) Greuter (*S. rupestris* Waldst. & Kit.) - cliffs slopes, screes - C

Senecio vulgaris L. (incl. *S. vulgaris* L. subsp. *denticulatus* (O. F. Müll.) P. D. Sell) - ruderal environments, fields, stony slopes - CC

Serratula tinctoria L. subsp. ***tinctoria*** - woods, humid meadows - R - Villavallelonga, Collelongo (Anzalone and Bazzichelli 1960).

Silybum marianum (L.) Gaertn. - uncultivated land, ruderal environments - PC

A ***Solidago canadensis*** L. - humid meadows - CAS - Lago Cardito (Conti and Minutillo 1998).

Solidago virgaurea L. subsp. ***virgaurea*** (incl. *S. pygmaea* G. Bertol.) - open woods, scrub, pastures - C

Sonchus asper (L.) Hill subsp. ***asper*** - ruderal environments - C

Sonchus bulbosus (L.) N. Kilian & Greuter subsp. ***bulbosus*** (*Aetheorhiza bulbosa* (L.) Cass.; *Leontodon bulbosus* L.) - thermophilous stony slopes - R - Monte Castelnuovo! (Conti 1992), Gole del Sagittario! (Conti 1995; Conti and Tinti 2012).

Sonchus oleraceus L. - ruderal environments - C

Sonchus tenerrimus L. - walls - PC

A ***Symphyotrichum novi-belgii*** (L.) G. L. Nesom (*Aster novi-belgii* L.; *A. laevigatus* Lam.; *A. novi-belgii* L. subsp. *laevigatus* (Lam.) Thell.) - humid environments - CAS - near S. Biagio Saracinisco (Conti 1995), Bisegna! (from a specimen collected by Minutillo in APP)

A ***Tanacetum balsamita*** L. (*Balsamita major* Desf.; *Chrysanthemum balsamita* (L.) Baill.) - ruderal environments - NAT

Tanacetum corymbosum (L.) Sch.Bip. subsp. ***achilleae*** (L.) Greuter (*Chrysanthemum achilleae* L.; *Pyrethrum tenuifolium* Willd.) - open woods and margins - C

Tanacetum parthenium (L.) Sch.Bip. (*Matricaria parthenium* L.) - open woods and margins - PC

Tanacetum vulgare L. subsp. ***vulgare*** - arid uncultivated land - PC

E ***Taraxacum glaciale*** É. Huet & A. Huet ex Hand.-Mazz. - snowbed meadows - C

Taraxacum sect. ***Alpina*** G. E. Haglund (incl. *T. apenninum* (Ten.) DC.) - montane pastures - C

Taraxacum sect. ***Erythrosperma*** (H. Lindb.) Dahlst. (incl. *T. laevigatum* (Willd.) DC.; incl. *T. fulvum* group) - ruderal environments, arid pastures - C

Taraxacum sect. ***Palustria*** (H. Lindb.) Dahlst. - marshy environments - PC

Taraxacum sect.***Taraxacum*** (incl. *T. officinale* group; *T.* sect. *Ruderalia* Kirschner, H. Øllg. & Štěpánek) - ruderal environments, fertilized meadows - CC

Tephroseris integrifolia (L.) Holub subsp. *capitata* (Wahlenb.) B. Nord. (*Cineraria capitata* Wahlenb.; *Senecio capitatus* (Wahlenb.) Steud.; *S. aurantiacus* auct. Fl. Ital.) - NC - generically recorded for the Park (Sipari 1926 quotes Pirotta; Lusina 1954).

Tephroseris integrifolia (L.) Holub subsp. ***integrifolia*** (*Senecio integrifolius* (L.) Clairv. subsp. *integrifolius*) - montane and subalpine meadows - C

E *Tolpis virgata* (Desf.) Bertol. subsp. *grandiflora* (Ten.) Arcang. (*T. grandiflora* Ten.) - NC - Picinisco (Tenore 1835; Tenore and Gussone 1842).

Tragopogon crocifolius L. subsp. ***crocifolius*** - arid meadows - PC

Tragopogon orientalis L. (*T. pratensis* L. subsp. *orientalis* (L.) Čelak.) - glades, meadows - PC

E ***Tragopogon porrifolius*** L. subsp. ***eriospermus*** (Ten.) Greuter (*T. eriospermus* Ten.) - meadows - PC

Tragopogon porrifolius L. subsp. ***porrifolius*** (incl. *T. australis* Jord.; incl. *T. porrifolius* L. subsp. *australis* (Jord.) Nyman) - arid meadows - C

Tragopogon pratensis L. - meadows - PC

Tragopogon samaritanii Heldr. & Sart. ex Boiss. (*T. crocifolius* L. subsp. *samaritani* (Heldr. & Sart. ex Boiss.) I. B. K. Richards) - arid meadows - PC - the records of *T. crocifolius* var. *nebrodensis* (Anzalone and Bazzichelli 1960) are to be referred to this species (Anzalone and Veri 1975).

Tripleurospermum inodorum (L.) Sch.Bip. (*Matricaria inodora* L.; *M. perforata* Mérat; *Tripleurospermum perforatum* (Mérat) Laínz) - ruderal environments - R - near Scanno (Anzalone 1961); Vallechiara near Pescasseroli (Conti 1998).

Tussilago farfara L. - humid clayey uncultivated land - CC

Urospermum dalechampii (L.) F. W. Schmidt (*Tragopogon dalechampii* L.) - arid fields and uncultivated land - PC

Urospermum picroides (L.) Scop. ex F. W. Schmidt (*Tragopogon picroides* L.) - uncultivated land, ruderal environments - PC

A ***Xanthium orientale*** L. subsp. ***italicum*** (Moretti) Greuter (*X. italicum* Moretti; *X. strumarium* L. subsp. *italicum* (Moretti) D. Löve) - arid uncultivated nitrified land - NAT

A ***Xanthium spinosum*** L. - arid uncultivated nitrified land - NAT

A *Xeranthemum annuum* L. - CAS - Fendidoro at Picinisco (Tenore 1835).

Xeranthemum cylindraceum Sm. (*X. foetidum* auct. Fl. Ital.) - arid meadows - C

Xeranthemum inapertum (L.) Mill. (*X. foetidum* Moench, nom. illeg.) - arid meadows - C

B

Balsaminaceae

A ***Impatiens balfourii*** Hook. f. - uncultivated land - NAT - Vallone del Lacerno! (Conti and Minutillo 1998), Val Canneto (Settefrati) (Petriglia 2004).

Impatiens noli-tangere L. - glades in cool woods - PC

Berberis vulgaris L. subsp. ***vulgaris*** - margins of woods, scrub - C

Betulaceae

A? ***Alnus cordata*** (Loisel.) Loisel. (*Betula cordata* Loisel.) - cool woods - NAT? - Camosciara (Conti 1995, 1998).

Alnus glutinosa (L.) Gaertn. (*Betula alnus* L. var. *glutinosa* L.; *B. glutinosa* (L) L.) - watercourses - R - Lago di Barrea (Spada 1979), Sagittario (Pirone 1995), Mainarde! (Conti 1995).

Betula pendula Roth - cool and stony glades generally in the *Fagus sylvatica* woods or at the upper limit of the tree vegetation - R - Coppo Oscuro di Barrea! (Bortolotti 1965; Bruno and Bazzichelli 1966), Valle Fredda! (Conti 1992), Vallone della Terratta! (Conti et al. 2002).

Carpinus betulus L. - cool woods - PC

Carpinus orientalis Mill. subsp. ***orientalis*** - mixed woods - PC

Corylus avellana L. - woods - C

Ostrya carpinifolia Scop. - woods - CC

F. Conti, F. Bartolucci, *The Vascular Flora of the National Park of Abruzzo, Lazio and Molise (Central Italy)*, Geobotany Studies, DOI 10.1007/978-3-319-09701-5_8

Blechnaceae

Blechnum spicant (L.) Roth. (*Osmunda spicant* L.) - H - Circumbor. - NC - Monte Forcellone at the Forestella (Tenore and Gussone 1842).

Boraginaceae

Anchusa azurea Mill. (*A. italica* Retz.) - uncultivated land, roadsides, arid pastures to the submontane belt - C

Anchusa officinalis L. - uncultivated land, roadsides, stony slopes to the montane belt - PC

Anchusella cretica (Mill.) Bigazzi, E. Nardi & Selvi (*Anchusa cretica* Mill.; *Lycopsis variegata* auct. Fl. Ital.) - NC - at Molino near S. Biagio (Zodda 1931).

Asperugo procumbens L. - uncultivated land, ruderal environments - PC

Borago officinalis L. - fields, ruderal environments - C

Buglossoides arvensis (L.) I. M. Johnst. subsp. ***arvensis*** - uncultivated land, arid pastures - C

Buglossoides incrassata (Guss.) I. M. Johnst. (*Lithospermum gasparrinii* Heldr. ex Guss.; *L. incrassatum* Guss.; *Buglossoides arvensis* (L.) I. M. Johnston subsp. *gasparrinii* (Heldr. ex Guss.) R. Fernandes; *B. gasparrinii* (Heldr. ex Guss.) Pignatti) - NC - Villavallelonga (Fiori et al. 1907). It has been observed near Carrito outside the Park but close to the buffer external zone.

Buglossoides purpurocaerulea (L.) I. M. Johnst. (*Lithospermum purpurocaeruleum* L.) - open woods - C

Cerinthe major L. subsp. *major* (*C. gymnandra* Gasp.; *C. major* L. subsp. *gymnandra* (Gasp.) Rouy) - NC - Mainarde (Zodda 1931).

Cerinthe minor L. subsp. ***auriculata*** (Ten.) Domac (*C. auriculata* Ten.) - *Fagus sylvatica* woods glades, scrub - PC

E ***Cynoglossum apenninum*** L. (*Solenanthus apenninus* Fisch. & C. A. Mey.) (Figs. 9 in chapter "Vegetation Features" and 1) - *Fagus sylvatica* wood glades, montane pastures - C

Cynoglossum columnae Ten. - arid and stony pastures - C

Cynoglossum creticum Mill. - uncultivated land - PC

E ***Cynoglossum magellense*** Ten. (Fig. 2) - stony pastures - C

Cynoglossum montanum L. - uncultivated areas, stony pastures, margins of woods - PC

Cynoglossum officinale L. - uncultivated areas, stony pastures, margins of woods - PC

Cynoglottis barrelieri (All.) Vural & Kit Tan subsp. ***barrelieri*** (*Buglossum barrelieri* All.; *Anchusa barrelieri* (All.) Vitman) - stony pastures, margins of woods, roadsides from the hill to the montane belt - CC

Echium italicum L. subsp. ***italicum*** - uncultivated areas, arid pastures - PC

Echium plantagineum L. - NC - Picinisco (Tenore and Gussone 1842).

Fig. 1 *Cynoglossum apenninum* (Photo by F. Conti)

Fig. 2 *Cynoglossum magellense* (Photo by F. Conti)

Echium vulgare L. subsp. ***pustulatum*** (Sm.) Em. Schmid & Gams (*E. pustulatum* Sm.) - arid pastures, ruderal environments - C

Echium vulgare L. subsp. ***vulgare*** - arid pastures, ruderal environments - PC

Heliotropium europaeum L. - ruderal environments, fields, arid uncultivated areas - C

Lappula squarrosa (Retz.) Dumort. (*Myosotis squarrosa* Retz.; *Lappula echinata* Gilib., des. inval.; *Echinospermum lappula* (L.) Lehm.) - arid uncultivated land, stony pastures - R - S. Donato on Monte Croce (Terracciano 1878 sub *Rochelia lappula* Rom. et Schultz), Fossato near Villavallelonga (Grande 1904), Gioia Vecchio (Anzalone and Bazzichelli 1960).

Lithospermum officinale L. - cool meadows, hedges - RR - Valle di Canneto (Anzalone and Bazzichelli 1960).

Myosotis arvensis (L.) Hill subsp. ***arvensis*** (*M. scorpioides* L. var. *arvensis* L.) - arid pastures, fields, uncultivated land - C

E ***Myosotis decumbens*** Host subsp. ***florentina*** Grau - glades, margins of woods - PC

E ***Myosotis graui*** Selvi (*M. ambigens* auct. Fl. Ital.) - montane pastures, stony slopes - CC

Myosotis incrassata Guss. - arid pastures - PC

Myosotis laxa Lehm. subsp. ***caespitosa*** (Schultz) Hyl. ex Nordh. (*M. caespitosa* Schultz) - marshy meadows - R - Montenero Val Cocchiara! (Corbetta and Pirone 1989), Val Fondillo!

Myosotis nemorosa *Besser* - woods and humid meadows - RR - Piana di Pescasseroli (Pedrotti et al. 1992).

Myosotis pusilla Loisel. - D - Villavallelonga (Grande 1904; Anzalone and Bazzichelli 1960).

Myosotis ramosissima Rochel subsp. ***ramosissima*** - arid pastures, uncultivated land - C

Myosotis scorpioides L. subsp. ***scorpioides*** - marshy meadows - PC

Myosotis stricta Link ex Roem. & Schult. - arid pastures - RR - Le Forme! (Conti 1995).

Myosotis sylvatica Hoffm. subsp. ***cyanea*** (Hayek) Vestergren (*M. cyanea* (Hayek) Domin; *M. sylvatica* f. *cyanea* Hayek) - woods, scrub - C - The records of *M. sylvatica* are to be referred to this taxon.

Neatostema apulum (L.) I. M. Johnst. - NC - S. Donato Val di Comino at Monte Croce (Terracciano 1878), between Pescasseroli and Barrea (Vaccari and Wilczek 1940).

E ***Onosma echioides*** (L.) L. subsp. ***echioides*** - arid and stony slopes - C

E ***Pulmonaria hirta*** L. subsp. ***apennina*** (Cristof. & Puppi) Peruzzi (*Pulmonaria apennina* Cristof. & Puppi) - woods - C - Probably are to be referred here the records of *P. vallarsae* and *P. officinalis*.

Symphytum officinale L. - humid meadows - RR - Lago Cardito (Conti and Minutillo 1998).

Symphytum tuberosum L. subsp. ***angustifolium*** (A. Kern.) Nyman (*S. tuberosum* L. subsp. *nodosum* (Schur) Soó) - woods - C - The record of *Symphytum bulbosum* K. F. Schimp. (Conti 1995) has to be referred to this taxon.

Brassicaceae

Aethionema saxatile (L.) R. Br. subsp. ***saxatile*** - stony slopes - C

Fig. 3 *Arabis collina* subsp. *rosea* (Photo by F. Conti)

Alliaria petiolata (M. Bieb.) Cavara and Grande (*Arabis petiolata* M. Bieb.) - ruderal environments - C

Alyssoides utriculata (L.) Medik. (*A. utriculata* (L.) Moench subsp. *graeca* (Boiss.) Jáv.; *Alyssum utriculatum* L.; *Vesicaria graeca* Boiss.) - cliffs, arid stony slopes - PC

Alyssum alyssoides (L.) L. (*Clypeola alyssoides* L.) - arid stony meadows - C

Alyssum cuneifolium Ten. subsp. *cuneifolium* - D - Val Fondillo (Falqui 1899), Settefrati at Pietrorosiello and Monte Maro (Terracciano 1878). Most probably to be excluded.

E ***Alyssum diffusum*** Ten. subsp. ***diffusum*** - stony pastures - C

Alyssum simplex Rudolphi (*A. campestre* auct. Fl. Ital.; *A. minus* Rothm., nom. illeg.) - arid meadows - PC

Arabidopsis thaliana (L.) Heynh. (*Arabis thaliana* L.) - fields, uncultivated land - PC

Arabis alpina L. subsp. ***caucasica*** (Willd.) Briq. (*A. caucasica* Willd.) - screes, shady screes, cliffs - C

Arabis auriculata Lam. (*A. recta* Vill.) - stony meadows - PC

Arabis bellidifolia Crantz subsp. ***stellulata*** (Bertol.) Greuter & Burdet (*A. pumila* Jacq. subsp. *stellulata* (Bertol.) Nyman; *A. stellulata* Bertol.) - montane cliffs and highest parts - PC

Arabis collina Ten. subsp. ***collina*** - cliffs, walls, scrub - C

Fig. 4 *Aubrieta columnae* subsp. *columnae* (Photo by F. Conti)

E ***Arabis collina*** Ten. subsp. ***rosea*** (DC.) Minuto (*A. rosea* DC.) (Fig. 3) - cliffs, walls - PC

Arabis hirsuta (L.) Scop. (*Turritis hirsuta* L.) - stony slopes, scrub - C

Arabis pauciflora (Grimm) Garcke (*A. brassica* (Leers) Rauschert; *Brassica alpina* L.; *Fourraea alpina* (L.) Greuter & Burdet) - *Fagus sylvatica* woods, scrub, hedges - PC

Arabis sagittata (Bertol.) DC. (*Turritis sagittata* Bertol.) - arid meadows, hedges, open woods - C

Arabis surculosa N. Terracc. (*A. serpillifolia* Vill. subsp. *nivalis* (Guss.) B. M. G. Jones) - snowbed meadows - PC

Arabis verna (L.) R. Br. (*Hesperis verna* L.) - cliffs, scrub - PC

E ***Aubrieta columnae*** Guss. subsp. ***columnae*** (Fig. 4) - cliffs, seldom on walls - PC

Barbarea bracteosa Guss. - humid environments - C

Barbarea stricta Andrz. - humid environments - R - Villetta Barrea (Anzalone and Bazzichelli 1960), Piana di Pescasseroli (Pedrotti et al. 1992), Val Canneto (Conti and Minutillo 1998).

****Barbarea vulgaris*** R. Br. subsp. ***arcuata*** (Opiz) Hayek (*Erysimum arcuatum* Opiz) - humid environments - RR - between S. Gennaro and Valleporcina!

Barbarea vulgaris R. Br. subsp. ***vulgaris*** - humid environments - PC

Biscutella cichoriifolia Loisel. - NC - generically recorded for the Park (Sipari 1926 quotes Pirotta; Lusina 1954).

E ***Biscutella laevigata*** L. subsp. ***australis*** Raffaelli and Baldoin - screes, stony pastures - PC

Biscutella laevigata L. subsp. ***laevigata*** - screes, stony pastures - C

Brassica gravinae Ten. - stony slopes - PC

A ***Brassica napus*** L. subsp. ***napus*** - uncultivated land - CAS

A ***Brassica rapa*** L. subsp. ***campestris*** (L.) Clapham - uncultivated land - CAS

Bunias erucago L. - ruderal environments, uncultivated land, stony slopes - C

Calepina irregularis (Asso) Thell. (*Myagrum irregulare* Asso) - uncultivated land - PC

Camelina sativa (L.) Crantz (*Myagrum sativum* L.; *Camelina sativa* var. *pilosa* DC.; *C. sativa* subsp. *pilosa* (DC.) N. W. Zinger) - NC - between Gioia Vecchio and Pescasseroli, Valico di Pantano (Vaccari and Wilczek 1940).

Capsella bursa-pastoris (L.) Medik. subsp. ***bursa-pastoris*** (*Thlaspi bursa-pastoris* L.) - uncultivated land, ruderal environments - C

Capsella rubella Reut. - uncultivated land, ruderal environments - R - generically recorded for the Park (Bazzichelli and Furnari 1970).

Cardamine amporitana Sennen & Pau (*C. amara* L. subsp. *grandifolia* Arcang.; *C. raphanifolia* p.p.) - humid environments - C

Cardamine bulbifera (L.) Crantz (*Dentaria bulbifera* L.) - *Fagus sylvatica* woods - CC

Cardamine chelidonia L. - *Fagus sylvatica* woods and cool woods - C

Cardamine enneaphyllos (L.) Crantz (*Dentaria enneaphyllos* L.) - *Fagus sylvatica* woods - C

Cardamine graeca L. - stony slopes in mixed woods and low *Fagus sylvatica* woods - C

Cardamine heptaphylla (Vill.) O. E. Schulz (*Dentaria heptaphylla* Vill.; *D. pinnata* Lam.) - NC - Picinisco in Valle dei Treconfini (Terracciano 1873).

Cardamine hirsuta L. - ruderal environments, scrub - CC

Cardamine impatiens L. subsp. ***impatiens*** - humid shady areas - C

Cardamine kitaibelii Bech. (*Dentaria polyphylla* Waldst. Kit.; *C. polyphylla* (Waldst. Kit.) O. E. Schulz) - *Fagus sylvatica* woods - C

Clypeola jonthlaspi L. subsp. ***jonthlaspi*** - arid meadows, stony slopes - PC

Clypeola jonthlaspi L. subsp. ***microcarpa*** (Moris) Arcang. (*C. microcarpa* Moris) - arid meadows, stony slopes - RR - Olmo di Bobbi! (Anzalone 1962).

Conringia austriaca (Jacq.) Sweet (*Brassica austriaca* Jacq.) - fields - R - Bisegna, tra Settefrati e la Valle di Canneto (Anzalone and Bazzichelli 1960).

Conringia orientalis (L.) Dumort. (*Brassica orientalis* L.) - fields - R - Villavallelonga (Grande 1904), Sperone!

Descurainia sophia (L.) Webb ex Prantl (*Sisymbrium sophia* L.) - ruderal environments - PC

Diplotaxis erucoides (L.) DC. subsp. ***erucoides*** (*Sinapis erucoides* L.) - cultivated and uncultivated land - C

Diplotaxis muralis (L.) DC. (*Sisymbrium murale* L.) - ruderal environments, uncultivated land - PC

Diplotaxis tenuifolia (L.) DC. (*Sisymbrium tenuifolium* L.) - walls, arid uncultivated land - C

Diplotaxis viminea (L.) DC. (*Sisymbrium vimineum* L.) - NC - Picinisco (Tenore and Gussone 1842).

Draba aizoides L. subsp. ***aizoides*** - cliffs, stony pastures - CC

Draba verna L. subsp. ***praecox*** (Steven) Rouy & Foucaud (*D. praecox* Steven; *Erophila praecox* (Steven) DC.; *E. verna* (L.) DC. subsp. *praecox* (Steven) Walp.) - arid uncultivated land - PC

Draba verna L. subsp. ***spathulata*** (Láng) Rouy & Foucaud (*D. spathulata* Láng; *Erophila spathulata* Láng; *E. verna* (L.) DC. subsp. *spathulata* (Láng) Vollm.) - arid uncultivated land - PC

Drabella muralis (L.) Fourr. (*Draba muralis* L.) - uncultivated land, walls - C

Eruca vesicaria (L.) Cav. (*Brassica longirostris* Uechtr.; *E. vesicaria* (L.) Cav. subsp. *longirostris* (Uechtr.) Rouy; *E. vesicaria* (L.) Cav. subsp. *sativa* (Mill.) Thell.; *Vesicaria sativa* Mill.) - ruderal environments - PC

A ***Erysimum cheiri*** (L.) Crantz (*Cheiranthus cheiri* L.) - walls - NAT

E ***Erysimum majellense*** Polatschek - stony slopes - PC

E ***Erysimum pseudorhaeticum*** Polatschek - stony slopes and stony pastures - CC

Fibigia clypeata (L.) Medik. (*Alyssum clypeatum* L.) - arid and stony slopes in the low - montane belt up to 1,200 m - PC

Hesperis laciniata All. subsp. ***laciniata*** - stony slopes, screes - C

Hesperis matronalis L. subsp. ***matronalis*** - cool *Fagus sylvatica* woods at their lower limit, on rocky substratum - PC

Hornungia alpina (L.) O. Appel subsp. ***alpina*** - cliffs, grassy ledges - RR - Monte Coppella (Petriccione 1988).

Hornungia petraea (L.) Rchb. subsp. ***petraea*** (*Lepidium petraeum* L.) - cliffs and arid meadows - PC

Iberis saxatilis L. subsp. ***saxatilis*** - stony areas, mountain tops - PC

Iberis umbellata L. - arid grassy uncultivated land - R - Valle di Canneto (Terracciano 1878), Colle della Regina (Conti 1995).

Iberis violacea R. Br. (*I. pruitii* Tineo; *I. carnosa* auct. Fl. Ital.) - arid pastures - PC

Isatis apennina Ten. ex Grande (*I. allionii* P. W. Ball.) - screes - RR - Monte e Serra Cappella (Petriccione 1988), Colle della Monna (Griebl 2010).

A ***Isatis tinctoria*** L. subsp. ***tinctoria*** (*I. canescens* DC.; *I. tinctoria* L. subsp. *canescens* (DC.) Arcang.) - arid uncultivated land - INV

Kernera saxatilis (L.) Sweet subsp. ***saxatilis*** (*Cochlearia saxatilis* L.) - cliffs - PC

Lepidium campestre (L.) R. Br. (*Thlaspi campestre* L.) - uncultivated land, ruins - C

Lepidium coronopus (L.) Al-Shehbaz (*L. squamatum* Forssk.; *Coronopus squamatus* (Forssk.) Asch.) - NC - Villavallelonga (Grande 1904).

Lepidium draba L. subsp. ***draba*** (*Cardaria draba* (L.) Desv.) - ruderal environments - C

Lepidium graminifolium L. subsp. ***graminifolium*** (*L. graminifolium* subsp. *suffruticosum* (L.) P. Monts.; *L. suffruticosum* L.) - ruderal environments - RR - Monte della Rocchetta near Colli al Volturno! (Conti 1995).

Lunaria annua L. (incl. *L. annua* subsp. *pachyrhiza* (Borbás) Hayek; *L. pachyrhiza* Borbás) - humid stony slopes - R - Pizzone (Conti 1995), Gole del Sagittario! (Conti and Tinti 2012).

Lunaria rediviva L. - gorges - C

Matthiola fruticulosa (L.) Maire subsp. *fruticulosa* - it has been observed near Colle Truscino outside the Park but close to the buffer external zone.

Microthlaspi perfoliatum (L.) F. K. Mey. (*Thlaspi perfoliatum* L.) - fields, uncultivated land - C

Fig. 5 *Noccaea stylosa* (Photo by F. Conti)

Myagrum perfoliatum L. - uncultivated land - R - Gioia Vecchio (Vaccari and Wilczek 1940), Val Cocchiara!

Nasturtium officinale R. Br. subsp. ***officinale*** - sources, riversides - C

Neslia paniculata (L.) Desv. subsp. ***thracica*** (Velen.) Bornm. (*N. apiculata* Fisch., C. A. Mey. & Avé-Lall.) - fields - PC

Noccaea brachypetala (Jord.) F. K. Mey. subsp. ***brachypetala*** (*Thlaspi brachypetalum* Jord.) - arid pastures - PC

Noccaea praecox (Wulfen) F. K. Mey. (*Thlaspi praecox* Wulfen) - arid pastures - C

E ***Noccaea stylosa*** (Ten.) Rchb. (*Iberis stylosa* Ten.; *Thlaspi stylosum* (Ten.) Mutel) (Fig. 5) - pebbly areas subjected to long periods of snow - PC

Pseudoturritis turrita (L.) Al - Shehbaz (*Arabis turrita* L.) - H - Medit. - stony and open woods - C

Raphanus raphanistrum L. subsp. ***landra*** (Moretti ex DC.) Bonnier and Layens (*R. landra* Moretti ex DC.; *R. maritimus* Sm.; *R. raphanistrum* subsp. *maritimus* (Sm.) Thell.) - ruderal environments - RR - Picinisco (along the way to Prati di Mezzo) (Petriglia 2004).

Raphanus raphanistrum L. subsp. ***raphanistrum*** (*R. raphanistrum* subsp. *microcarpus* (Lange) Thell.; *R. raphanistrum* var. *microcarpus* Lange) - ruderal environments - RR - Opi (Conti 1995).

A *Rapistrum perenne* (L.) All. (*Myagrum perenne* L.) - CAS - Valle dell'Acquaro over Fendidoro (Tenore and Gussone 1842).

Rapistrum rugosum (L.) All. (*Myagrum rugosum* L.; *R. rugosum* (L.) All. subsp. *linnaeanum* Rouy & Foucaud; *R. rugosum* subsp. *orientale* (L.) Arcang.) - fields, ruderal environments - PC

Rorippa palustris (L.) Besser (*Sisymbrium amphibium* L. var. *palustre* L.) - NC - territory of Villavallelonga (Grande 1904).

Rorippa sylvestris (L.) Besser subsp. ***sylvestris*** (*Sisymbrium sylvestre* L.) - humid environments - C

A *Sinapis alba* L. subsp. *alba* - CAS - Mainarde (Zodda 1931).

Sinapis arvensis L. subsp. ***arvensis*** - fields, uncultivated land - RR - Pescasseroli (Anzalone and Bazzichelli 1960 as *Brassica arvensis* L. *orientalis* Fiori).

****Sisymbrium altissimum*** L. - ruderal environments - RR - between Collelongo and Trasacco!

Sisymbrium irio L. - ruderal environments - R - Monte Castelnuovo near the village! (Conti 1995), Colle della Monna (Griebl 2010).

Sisymbrium officinale (L.) Scop. (*Erysimum officinale* L.) - ruderal environments - C

Sisymbrium orientale L. subsp. ***orientale*** - ruderal environments - C

Thlaspi alliaceum L. - fields, uncultivated land - PC

Thlaspi arvense L. - fields, ruderal environments - PC

Turritis glabra L. (*Arabis pseudoturritis* Boiss. & Heldr.; *A. glabra* (L.) Bernh.) - woods - PC

Buxaceae

Buxus sempervirens L. - thermophilous woods, scrub and stony supramediterranean garigue to 1,200 m - R - Lecce nei Marsi! (Anzalone and Bazzichelli 1960).

C

Campanulaceae

Campanula bononiensis L. - open woods, glades - PC
Campanula cochleariifolia Lam. - cliffs at high altitude - PC
Campanula erinus L. - cliffs and walls - PC
Campanula foliosa Ten. - pastures and *Fagus sylvatica* wood glades - PC
E ***Campanula fragilis*** Cirillo subsp. ***cavolinii*** (Ten.) Damboldt (*C. cavolinii* Ten.) (Fig. 1) - cliffs - PC
Campanula glomerata L. (incl. *C. glomerata* L. subsp. *cervicarioides* (Schult.) Arcang.; incl. *C. glomerata* L. subsp. *elliptica* (Kit. ex Schult.) Jan; incl. *C. glomerata* L. subsp. *farinosa* (Andrz.) Kirschl.; incl. *C. glomerata* L. subsp. *serotina* (Wettst.) O. Schwartz) - stony pastures - C
Campanula latifolia L. - open woods and margins - PC
E ***Campanula micrantha*** Bertol. (*C. apennina* Podlech) - stony pastures - C
Campanula persicifolia L. subsp. ***persicifolia*** - open woods, scrub - C
Campanula rapunculus L. - uncultivated land and arid pastures - C
Campanula rotundifolia L. s.l. - D - the records of this species (Tenore 1835; Tenore and Gussone 1842; Terracciano 1872; Anzalone and Bazzichelli 1960) are probably to be referred to other taxa of the subsect. *Heterophylla* (*C. micrantha*, *C. scheuchzeri*, *C. tanfanii*).
Campanula scheuchzeri Vill. subsp. ***scheuchzeri*** - pastures, stony slopes - C
E ***Campanula tanfanii*** Podlech - cliffs - C
Campanula trachelium L. subsp. ***trachelium*** - glades, open woods - C
Edraianthus graminifolius (L.) A. DC. subsp. ***graminifolius*** (*E. graminifolius* (L.) DC. subsp. *apenninus* Lakusic) - cliffs, stony slopes - C
Jasione montana L. (*J. echinata* Boiss. & Reut.; *J. montana* L. subsp. *echinata* (Boiss. & Reuter) Nyman) - arid sandy uncultivated land - RR - Lago di Barrea (Hennecke and Hennecke 1999).

F. Conti, F. Bartolucci, *The Vascular Flora of the National Park of Abruzzo, Lazio and Molise (Central Italy)*, Geobotany Studies, DOI 10.1007/978-3-319-09701-5_9

Fig. 1 *Campanula fragilis* Cirillo subsp. *cavolinii* (Photo by F. Conti)

Legousia falcata (Ten.) Janch. subsp. ***falcata*** - fields, arid uncultivated land - PC
Legousia hybrida (L.) Delarbre - fields - PC
Legousia speculum-veneris (L.) Chaix - fields - C
Phyteuma hemisphaericum L. - NC - Valle dell'Acquaro (Tenore 1835), between Monte Cavallo and la Parruccia (Zodda 1931).
Phyteuma orbiculare L. - pastures, stony slopes - C

Cannabaceae

Celtis australis L. subsp. ***australis*** - stony slopes and thermophilous woods - R - Monte Serra Traversa (Spada 1979), Scanno (Viegi et al. 1990).
Humulus lupulus L. - watercourses - C

Caprifoliaceae

Centranthus ruber (L.) DC. subsp. ***ruber*** (*Valeriana rubra* L.) - thermophilous cliffs, walls - PC
Cephalaria leucantha (L.) Roem. & Schult. (*Scabiosa leucantha* L.) - arid meadows, stony slopes - C
Cephalaria transsylvanica (L.) Roem. & Schult. (*C. allionii* Strobl; *Scabiosa transsylvanica* L.) - NC - Picinisco (Tenore and Gussone 1842), near S. Giuseppe (Zodda 1931).
Dipsacus fullonum L. - humid uncultivated land, clayey fields - C
Knautia integrifolia (L.) Bertol. subsp. ***integrifolia*** - arid pastures - PC
Knautia timeroyi Jord. subsp. ***collina*** (Schübler & G. Martens) Breistr. (*K. collina* Jord.; *K. purpurea* (Vill.) Borbás; *Scabiosa arvensis* L. subsp. *collina* Schübler & G. Martens) - arid pastures - CC - the records of *K. sylvatica* and *K. arvensis* are to be referred to this taxon (Conti 1995).

Lomelosia argentea (L.) Greuter & Burdet (*Scabiosa argentea* L.) - arid meadows and garigue - R - over Gioia dei Marsi (Anzalone and Bazzichelli 1960 as *Scabiosa argentea* L. *alba* (Scop.)), presso Villavallelonga!

E ***Lomelosia crenata*** (Cirillo) Greuter & Burdet subsp. ***pseudisetensis*** (Lacaita) Greuter & Burdet (*Scabiosa pseudisetensis* (Lacaita) Pignatti) - detrital slopes - PC

Lomelosia graminifolia (L.) Greuter & Burdet subsp. ***graminifolia*** (*Scabiosa graminifolia* L.) - stony slopes - C

Lonicera alpigena L. subsp. ***alpigena*** - *Fagus sylvatica* woods, scrub - C

Lonicera caprifolium L. - mixed woods, maquis, hedges - PC

Lonicera etrusca Santi - thermophilous scrub, open woods, hedges - C

Lonicera implexa Aiton subsp. *implexa* - D - near Barrea (Hennecke and Hennecke 1999).

Lonicera xylosteum L. var. ***nigra*** Loisel. - woods of broadleaves, scrub - C

Scabiosa columbaria L. s.l. - arid pastures, margins of woods - CC - We are conducting a biosystematics review of this aggregate for the purpose of ascertaining the actual presence of *S. columbaria*s.s. south of the Alps and understand the identity of *S. portae*, *S. levieri*, *S. ceratophylla*, *S. pyrenaica*, *S. uniseta* and *S. magellensis*, all indicated for the central Apennines.

Scabiosa silenifolia Waldst. & Kit. (Fig. 2) - cliffs, stony mountain tops - PC

Sixalix atropurpurea (L.) Greuter & Burdet subsp. ***grandiflora*** (Scop.) Soldano & F. Conti (*Scabiosa maritima* L.; *Sixalix atropurpurea* (L.) Greuter & Burdet subsp. *maritima* (L.) Greuter & Burdet) - arid meadows, uncultivated land - C

A ***Symphoricarpos albus*** (L.) S. F. Blake (*Vaccinium album* L.) - hedges - CAS - Villetta Barrea (Viegi et al. 1990).

Valeriana montana L. - screes, humid stony slopes - C

Valeriana officinalis L. - meadows and humid woods - C

Valeriana saliunca All. (Fig. 3) - stony meadows at high altitude - RR - Monte Greco (Di Pietro et al. 2004).

Valeriana tripteris L. subsp. ***tripteris*** - screes, humid stony slopes - PC

Fig. 2 *Scabiosa silenifolia* (Photo by F. Conti)

Fig. 3 *Valeriana saliunca* (Photo by F. Conti)

Valeriana tuberosa L. - arid pastures - CC

Valeriana wallrothii Kreyer (*V. collina* Wallr.; *V. officinalis* L. subsp. *collina* (Wallr.) Nyman) - humid meadows - PC

Valerianella carinata Loisel. - fields, uncultivated land - C

Valerianella coronata (L.) DC. - uncultivated land, arid meadows - C

Valerianella dentata (L.) Pollich - arid meadows, uncultivated land - PC

Valerianella discoidea (L.) Loisel. - NC - between Gioia dei Marsi and Gioia Vecchio (Vaccari and Wilczek 1940).

Valerianella echinata (L.) DC. - fields, arid uncultivated land - RR - Campoli Appennino! (Terracciano 1874; Conti and Minutillo 2001).

Valerianella eriocarpa Desv. - arid meadows - R - slopes of Monte Mattone near Pizzone! (Conti 1995), between Collelongo and Trasacco!

Valerianella locusta (L.) Laterr. - fields, arid meadows - C

Valerianella muricata (Stev. ex M. Bieb.) J. W. Loudon - arid pastures - RR - Gole del Sagittario! (Conti 1998).

Valerianella puberula (Bertol. ex Guss.) DC. - NC - Picinisco, sorgente del Melfa (Tenore and Gussone 1842).

Valerianella pumila (L.) DC. - fields - RR - Villavallelonga (Anzalone and Bazzichelli 1960).

Caryophyllaceae

Agrostemma githago L. - fields - PC

E ***Arenaria bertolonii*** Fiori & Paol. - screes, shady stony slopes - C

Arenaria grandiflora L. subsp. ***grandiflora*** - stony slopes at high altitude - PC

Arenaria leptoclados (Rchb.) Guss. subsp. ***leptoclados*** (*A. serpyllifolia* L. var. *leptoclados* Rchb.; *A. serpyllifolia* L. subsp. *leptoclados* (Rchb.) Nyman; *A. serpyllifolia* L. subsp. *tenuior* (Mert. & W. D. J. Koch) Arcang.) - arid meadows and cliffs - PC

Arenaria serpyllifolia L. subsp. ***serpyllifolia*** - uncultivated land, arid and stony meadows - C

Atocion armeria (L.) Raf. (*Silene armeria* L.) - stony meadows - PC

Bufonia paniculata Dubois ex Delarbre - arid uncultivated land - R - Villavallelonga! (Grande 1924; Conti 1995), Collelongo (Anzalone and Bazzichelli 1960).

Cerastium arvense L. subsp. *arvense* (incl. *C. etruscum* Lacaita) - D - the records from the Park territory are to be referred probably to *C. arvense* subsp. *suffruticosum*.

Cerastium arvense L. subsp. ***suffruticosum*** (L.) Ces. (*C. suffruticosum* L.) - stony meadows and pastures - C

Cerastium brachypetalum Desp. ex Pers. subsp. ***roeseri*** (Boiss. & Heldr.) Nyman (*C. luridum* Guss.; *C. roeseri* Boiss. & Heldr.) - arid environments - C

Cerastium brachypetalum Desp. ex Pers. subsp. *tauricum* (Spreng.) Murb. (*C. tauricum* Spreng.) - D

Cerastium brachypetalum Desp. ex Pers. subsp. *tenoreanum* (Ser.) Soó & Jáv. (*C. tenoreanum* Ser.) - D - Colle della Monna (Griebl 2010).

Cerastium cerastoides (L.) Britton (*Stellaria cerastoides* L.) - snowbed meadows - PC

Cerastium glomeratum Thuill. - ruderal environments, arid meadows - PC

Cerastium holosteoides Fr. (*C. fontanum* Baumg. subsp. *vulgare* (Hartm.) Greuter & Burdet; *C. holosteoides* Fr. subsp. *triviale* (Link) Moschl; *C. triviale* Link; *C. vulgare* Hartm.) - humid meadows - C

Cerastium ligusticum Viv. - stony pastures - PC

Cerastium pumilum Curtis (incl. *C. pumilum* Curtis subsp. *glutinosum* (Fr.) Jalas; incl. *C. glutinosum* Fr.) - arid pastures - PC

E ***Cerastium scaranoi*** Ten. - arid stony slopes - PC

Cerastium semidecandrum L. subsp. ***semidecandrum*** - stony pastures - PC

Cerastium siculum Guss. - fields, arid pastures - R - Rocchetta Nuova (Conti 1995), Gole del Sagittario! (Conti and Tinti 2012).

Cerastium sylvaticum Waldst. & Kit. - humid environments - RR - Camosciara!, riversides of Sangro river near Alfedena! (Conti 1995).

E *Cerastium thomasii* Ten. - D - eastern tops of Monte Marsicano (Petriccione 1988).

E ***Cerastium tomentosum*** L. - stony pastures, screes - CC

Cucubalus baccifer L. - humid woods - R - near Alfedena! (Conti 1995), Pantanello!

Dianthus armeria L. subsp. ***armeria*** - pastures - PC

Dianthus barbatus L. subsp. ***compactus*** (Kit.) Heuff. (*D. compactus* Kit.) - bushy pastures - PC

E ***Dianthus brachycalyx*** Huet ex Bacch., Brullo, Casti & Giusso (Fig. 4) - stony slopes - C - some samples show a variability of characters that does not fit in this species and deserve further study.

E ***Dianthus carthusianorum*** L. subsp. ***tenorei*** (Lacaita) Pignatti (*D. carthusianorum* L. var. *tenorei* Lacaita) (Fig. 5) - stony meadows - C

E *Dianthus* cfr. *guliae* Janka (*D. ferrugineus* auct. Fl. Ital.) - D - Gole del Sagittario! (Conti and Tinti 2012), Gioia dei Marsi!, Frattura! The taxonomic identity of populations from Abruzzo need further studies.

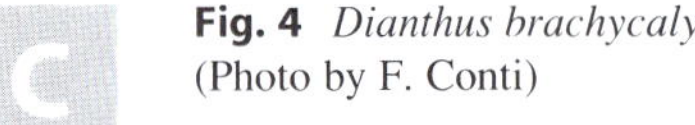

Fig. 4 *Dianthus brachycalyx* (Photo by F. Conti)

Dianthus ciliatus Guss. subsp. ***ciliatus*** - cliffs, stony termophilous slopes - PC

Dianthus deltoides L. subsp. ***deltoides*** - meadows - C

Dianthus hyssopifolius L. (incl. *D. marsicus* Ten.; incl. *D. sternbergii* Sieber ex Capelli subsp. *marsicus* (Ten.) Pignatti; *D. monspessulanus* L.; *D. monspeliacus* L.; incl. *D. waldsteinii* Sternb. subsp. *marsicus* (Ten.) Greuter & Burdet) - stony meadows - C

Dianthus longicaulis Ten. (*D. sylvestris* Wulfen subsp. *longicaulis* (Ten.) Greuter & Burdet; *D. caryophyllus* L. subsp. *longicaulis* (Ten.) Arcang.; *D. caryophyllus* L. var. *longicaulis* (Ten.) Fiori) - stony slopes - C

Drypis spinosa L. subsp. ***spinosa*** - screes - PC

Heliosperma pusillum (Waldst. & Kit.) Rchb. (*Silene pusilla* Waldst. & Kit.) - shady and wet cliffs - PC

Herniaria glabra L. subsp. ***glabra*** - arid environments and glades - PC

Herniaria glabra L. subsp. ***nebrodensis*** Jan ex Nyman (*Herniaria microcarpa* C. Presl ; *Sagina microcarpa* C. Presl) - stony pastures and high meadows - C

Herniaria incana Lam. - stony pastures - PC

Holosteum umbellatum L. subsp. ***umbellatum*** - arid uncultivated land - PC

E ***Mcneillia graminifolia*** (Ard.) Dillenb. & Kadereit subsp. ***rosanoi*** (Ten.) F. Conti, Bartolucci, Iamonico & Del Guacchio (*Arenaria rosanoi* Ten.; *Minuartia graminifolia* (Ard.) Jáv. subsp. *rosanoi* (Ten.) Mattf.) (4 in chapter

Fig. 5 *Dianthus carthusianorum* subsp. *tenorei* (Photo by F. Conti)

"Vegetation Features" and 6) - cliffs - PC - The nomenclature according to Bartolucci et al. (2014a).

Minuartia capillacea (All.) Graebn. (*Arenaria capillacea* All.) - cliffs, stony slopes - R - Monte Argatone!, Monte Terratta! (Grande 1914; Conti 1998).

E ***Minuartia glomerata*** (M. Bieb.) Degen subsp. ***trichocalycina*** (Ten. & Guss.) F. Conti (*Arenaria trichocalycina* Ten. & Guss.; *M. trichocalycina* (Ten. & Guss.) Grande) (Fig. 7) - arid and stony slopes from 850 to 1,300 m - PC

Moehringia muscosa L. - cool woods - PC

Moehringia trinervia (L.) Clairv. (*Arenaria trinervia* L.) - cool woods - PC

Paronychia kapela (Hacq.) A. Kern. subsp. ***kapela*** - stony cliffs and pastures - PC

Petrorhagia prolifera (L.) P.W. Ball & Heywood (*Dianthus prolifer* L.; *Tunica prolifera* (L.) Scop.) - arid meadows - PC

Petrorhagia saxifraga (L.) Link subsp. ***saxifraga*** - arid and stony slopes - C

Polycarpon tetraphyllum (L.) L. subsp. ***diphyllum*** (Cav.) O. Bolòs & Font Quer - ruderal environments - RR - Monte Mattone near Pizzone! (Conti 1995).

Polycarpon tetraphyllum (L.) L. subsp. ***tetraphyllum*** - fields and ruderal environments - R - Picinisco (Tenore and Gussone 1842), near S. Biagio (Zodda 1931 sub *P. tetraphyllum* fo. *verticillatum* Fenzl.), Colle della Monna (Griebl 2010).

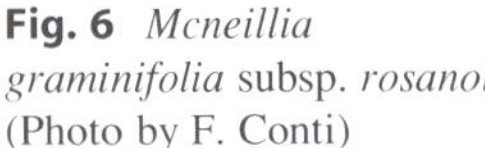

Fig. 6 *Mcneillia graminifolia* subsp. *rosanoi* (Photo by F. Conti)

Fig. 7 *Minuartia glomerata* subsp. *trichocalycina* (Photo by F. Conti)

Sabulina attica (Boiss. & Sprun.) Dillenb. & Kadereit (*Alsine attica* Boiss. & Spruner; *Minuartia verna* (L.) Hiern subsp. *attica* (Boiss. & Spruner) Graebn.) - stony slopes - C - The nomenclature according to Dillenberger and Kadereit (2014).

Sabulina glaucina (Dvořáková) Dillenb. & Kadereit (*Minuartia glaucina* Dvořáková; *Alsine verna* L. var. *collina* Neilr.; *Minuartia verna* (L.) Hiern subsp. *collina* (Neilr.) Domin) - stony slopes - C - The nomenclature according to Dillenberger and Kadereit (2014).

Sabulina mediterranea (Ledeb. ex Link) Rchb. (*Arenaria mediterranea* Ledeb. ex Link; *Minuartia mediterranea* (Ledeb. ex Link) K. Malý) - arid meadows - RR - Canneto! (Conti 1995). The nomenclature according to Dillenberger and Kadereit (2014).

Sabulina tenuifolia (L.) Rchb. subsp. ***tenuifolia*** (*Arenaria tenuifolia* L.; *Minuartia tenuifolia* (L.) Hiern, non Nees ex Mart., nom. illeg.; *Arenaria hybrida* Vill.; *Minuartia hybrida* (Vill.) Shischk. subsp. *hybrida*) - arid meadows - C - The nomenclature according to Dillenberger and Kadereit (2014).

Sabulina verna (L.) Rchb. subsp. ***verna*** (*Minuartia verna* (L.) Hiern subsp. *verna)* - stony slopes, high cliffs - PC - The nomenclature according to Dillenberger and Kadereit (2014).

Sagina apetala Ard. subsp. ***apetala*** - stony pastures - RR - Le Forme! (Conti 1995).

Sagina glabra (Willd.) Fenzl (*Spergula glabra* Willd.) - meadows and upland plains at high altitude - C

Sagina saginoides (L.) H. Karst. subsp. *saginoides* (*Spergula saginoides* L.) - D - Monte Forcellone alla Forestella (Tenore and Gussone 1842), Zaffineto (Terracciano 1873).

Sagina subulata (Sw.) C. Presl (*Spergula subulata* Sw.) - NC - Pescosolido (Terracciano 1873), Pescasseroli (Anzalone and Bazzichelli 1960 from a specimen collected by Grande).

Saponaria bellidifolia Sm. - cliffs, stony slopes - PC

Saponaria ocymoides L. subsp. ***ocymoides*** - stony slopes, screes - PC

Saponaria officinalis L. - humid uncultivated land - C

Scleranthus annuus L. - sterile soils in stony pastures - PC

Scleranthus polycarpos L. (*S. annuus* L. subsp. *polycarpos* (L.) Thell.) - sterile soils - PC

Scleranthus uncinatus Schur - sterile soils in pastures - PC

Scleranthus verticillatus Tausch (*S. annuus* L. subsp. *verticillatus* (Tausch) Arcang.; *S. polycarpos* L. subsp. *collinus* (Hornung) Pignatti) - sterile soils - PC

Silene acaulis (L.) Jacq. subsp. ***bryoides*** (Jord.) Nyman (*S. acaulis* (L.) Jacq. subsp. *exscapa* (All.) Braun-Blanq.; *S. bryoides* Jord.) - highest parts of the cliff environments - PC

Silene catholica (L.) W. T. Aiton (*Cucubalus catholicum* L.) - cool woods - PC

E ***Silene cattariniana*** Ferrarini & Cecchi (Fig. 8) - montane stony meadows - PC

Silene ciliata Pourr. subsp. ***graefferi*** (Guss.) Nyman (*S. graefferi* Guss.) - stony montane meadows - PC

Silene conica L. - arid meadows - PC

Silene dioica (L.) Clairv. (*Lychnis dioica* L.) - margins of *Fagus sylvatica* woods - PC

Silene flos-cuculi (L.) Clairv. (*Lychnis flos-cuculi* L.) - meadows - PC

Silene gallica L. (*S. lusitanica* L.) - NC - Picinisco (Tenore and Gussone 1842).

Silene inaperta L. - NC - Costa di Cavallaro at Monte Meta (Terracciano 1872).

Silene italica (L.) Pers. subsp. ***italica*** - scrub, open woods, stony slopes - CC

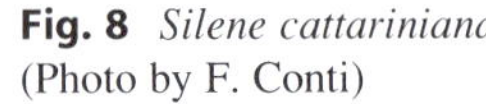

Fig. 8 *Silene cattariniana* (Photo by F. Conti)

Silene latifolia Poir. subsp. ***alba*** (Mill.) Greuter & Burdet (*Lychnis alba* Mill.; *S. alba* (Mill.) Krause) - ruderal environments and uncultivated areas - C

Silene latifolia Poir. subsp. ***latifolia*** - ruderal environments and uncultivated areas - C

Silene multicaulis Guss. subsp. ***multicaulis*** - stony slopes, cliffs - C

Silene nutans L. subsp. ***nutans*** - open woods, meadows - PC

Silene otites (L.) Wibel subsp. ***otites*** - arid meadows - C

Silene paradoxa L. - stony pastures - C

Silene pendula L. (Fig. 9) - stony slopes, uncultivated land - C

E ***Silene roemeri*** Friv. subsp. ***staminea*** (Bertol.) Nyman (*S. staminea* Bertol.) - alpine pastures - PC

Silene saxifraga L. - shady cliffs - PC

Silene viridiflora L. - woods - PC

Silene vulgaris (Moench) Garcke subsp. ***prostrata*** (Gaudin) Schinz & Thell. (*S. inflata* Sm. subsp. *prostrata* Gaudin; *S. uniflora* Roth subsp. *prostrata* (Gaudin) Chater & Walters) - screes - R - Monte Marsicano! (Bassani 1994 as *S. vulgaris* (Moench) Garcke subsp. *glareosa* (Jord.) Marsd. - Jet Turr.), Monte Cappella (from a specimen collected by Petriccione in the Herb. of the Park), Valle Cerasa!

Fig. 9 *Silene pendula* (Photo by F. Conti)

Silene vulgaris (Moench) Garcke subsp. ***tenoreana*** (Colla) Soldano & F. Conti (*S. tenoreana* Colla; *S. vulgaris* (Moench) Garcke subsp. *angustifolia* (Mill.) Hayek) - arid meadows and glades - PC

Silene vulgaris (Moench) Garcke subsp. ***vulgaris*** - arid meadows and glades - PC

Spergularia rubra (L.) J. Presl & C. Presl (*Arenaria rubra* L.) - meadows with *Nardus stricta* - RR - near Monte Forcellone! (Conti 1995).

Stellaria aquatica (L.) Scop. (*Cerastium aquaticum* L.; *Myosoton aquaticum* (L.) Moench) - NC - "lungo la Forma" (Grande 1904).

Stellaria graminea L. - humid meadows - PC

Stellaria holostea L. subsp. ***holostea*** - margins of woods - C

Stellaria media (L.) Vill. subsp. ***media*** - ruderal environments - CC

Stellaria nemorum L. subsp. ***montana*** (Pierrat) Berher (*S. montana* Pierrat; *S. nemorum* L. subsp. *glochidisperma* Murb.) - cool woods - C

Celastraceae

Euonymus europaeus L. - woods, margins of woods, hedges - C

Euonymus latifolius (L.) Mill. - cool woods - C

Euonymus verrucosus Scop. - cool woods, gorges - PC

Parnassia palustris L. subsp. ***palustris*** - humid meadows and wet cliffs - R - Camosciara! (Fiori 1927), Valle di Fondillo! (Anzalone and Bazzichelli 1960).

Cistaceae

Cistus creticus L. subsp. ***eriocephalus*** (Viv.) Greuter & Burdet (*C. eriocephalus* Viv.; *C. incanus* L.) - degraded garigue and maquis - PC

Cistus salviifolius L. - garigue - PC

Fumana ericifolia Wallr. (*F. ericoides* auct. Fl. Ital.; *F. ericoides* (Cav.) Gand. subsp. *montana* (Pomel) Güemes & Muñoz Garm.) - arid meadows - R - Monte Castelnuovo!, near the Abbazia of Castel S. Vincenzo! (Conti 1995).

Fumana procumbens (Dunal) Gren. & Godr. (*Cistus fumana* L.; *Fumana vulgaris* Spach; *Helianthemum fumana* (L.) Mill.; *H. procumbens* Dunal) - arid meadows - PC

Fumana thymifolia (L.) Spach ex Webb (*Cistus thymifolius* L.; *Helianthemum thymifolium* (L.) Pers.) - arid meadows - R - between Gioia dei Marsi and Gioia Vecchio! (Pirone and Tammaro 1997), near the Abbazia of Castel S. Vincenzo! (Conti 1995).

Helianthemum apenninum (L.) Mill. subsp. ***apenninum*** (*Cistus apenninus* L.) - stony pastures, scrub - CC

Helianthemum nummularium (L.) Mill. subsp. ***glabrum*** (W. D. J. Koch) Wilczek (*H. nitidum* Clementi; *H. nummularium* (L.) Mill. var. *glabrum* W. D. J. Koch) - RR - generically recorded for the Park (Anzalone and Bazzichelli 1960 sub *H. chamaecistus* Mill. *glabrum* (Kern.)).

Helianthemum nummularium (L.) Mill. subsp. ***grandiflorum*** (Scop.) Schinz & Thell. (*Cistus grandiflorus* Scop.; *H. grandiflorum* (Scop.) Lam.; *H. grandiflorum* (Scop.) DC.) - stony pastures, scrub - PC

Helianthemum nummularium (L.) Mill. subsp. ***nummularium*** - stony pastures, scrub - C

Helianthemum nummularium (L.) Mill. subsp. ***obscurum*** (Čelak.) Holub (*H. chamaecistus* Mill. subsp. *obscurum* Čelak.; *H. hyssopifolium* Ten.; *H. obscurum* Pers.; *H. ovatum* (Viv.) Dunal) - stony pastures, scrub - C

Helianthemum nummularium (L.) Mill. subsp. ***tomentosum*** (Scop.) Schinz & Thell. (*Cistus tomentosus* Scop.) - stony pastures, scrub - PC

Helianthemum oelandicum (L.) Dum. Cours. subsp. ***alpestre*** (Jacq.) Ces. (*Cistus alpestris* Jacq.; *H. alpestre* (Jacq.) DC.) - stony pastures, scrub - stony slopes - PC

Helianthemum oelandicum (L.) Dum. Cours. subsp. ***incanum*** (Willk.) G. López (*H. canum* (L.) Baumg.; *H. canum* (L.) Baumg. subsp. *canum*; *H. oelandicum* (L.) DC. subsp. *canum* (L.) Bonnier) - stony slopes - CC

Helianthemum oelandicum (L.) Dum. Cours. subsp. ***italicum*** (L.) Ces. (*Cistus italicus* L.; *H. italicum* (L.) Pers.) - stony slopes - PC

Helianthemum salicifolium (L.) Mill. (*Cistus salicifolius* L.) - arid meadows - PC

Fig. 10 *Colchicum bulbocodium* subsp. *versicolor* (Photo by F. Conti)

Colchicaceae

Colchicum alpinum DC. (*C. parvulum* Ten.; *C. alpinum* DC. subsp. *parvulum* (Ten.) Nyman) - cool meadows - C

Colchicum bulbocodium Ker Gawl. subsp. ***versicolor*** (Ker Gawl.) K. Perss. (*Bulbocodium vernum* L. subsp. *versicolor* (Ker Gawl.) K. Richt.; *B. versicolor* (Ker Gawl.) Spreng.; *Colchicum versicolor* Ker Gawl.) (Fig. 10) - arid pastures - RR - at Feudo di Sipari (Grande 1904), near Opi (Vitale in verb.).

Colchicum lusitanum Brot. (*C. actupii* Fridl.) - pastures - PC

E ***Colchicum neapolitanum*** (Ten.) Ten. (*C. autumnale* L. var. *napolitanum* Ten.; *C. multiflorum* auct. Fl. Ital.) - pastures - PC

Convolvulaceae

Calystegia sepium (L.) R. Br. subsp. ***sepium*** - hedges, humid woods - C

Convolvulus arvensis L. - fields, uncultivated land - C

Convolvulus cantabrica L. - arid meadows - CC

A ***Cuscuta campestris*** Yunck. (*C. gronovii* auct. Fl. Ital.; *C. suaveolens* auct. Fl. Ital. p.p.) - on *Xanthium* - NAT - the report of *C. cesattiana* Bertol. for Montenero Val Cocchiara (Conti and Minutillo 2001) is to be referred to this taxon.

*A ***Cuscuta epilinum*** Weihe - on weeds - CAS - Vallone d'Onofrio!

Cuscuta epithymum (L.) L. subsp. ***kotschyi*** (Des Moul.) Arcang. (*C. kotschyi* Des Moul.) - C

Cuscuta europaea L. - on *Urtica dioica*, *Sambucus ebulus*, etc. - C

Cuscuta planiflora Ten. - NC - Valle della Melfa (Terracciano 1873), Mainarde (Zodda 1931).

C

Cornaceae

Cornus mas L. - scrub, hedges - C
Cornus sanguinea L. subsp. ***hungarica*** (Kárpáti) Soó - hedges, scrub, woods - CC

Crassulaceae

Hylotelephium maximum (L.) Holub subsp. ***maximum*** (*Sedum maximum* (L.) Suter; *S. telephium* L. var. *maximum* L.) - walls and shady cliffs, screes - PC
Sedum acre L. (*S. acre* L. subsp. *neglectum* (Ten.) Rouy & Camus; *S. neglectum* Ten.) - walls, screes, stony slopes - CC
Sedum album L. subsp. ***micranthum*** (Bast. ex DC.) Syme (*S. micranthum* Bast. ex DC.) - walls and cliffs - CC
Sedum amplexicaule DC. subsp. ***tenuifolium*** (Sm.) Greuter (*S. tenuifolium* (Sm.) Strobl) - stony areas and cliffs - R - Villavallelonga (Grande 1924), between Gioia dei Marsi and Gioia Vecchio (Vaccari and Wilczek 1940).
Sedum atratum L. - stony meadows at high altitude, stony slopes - PC
Sedum caespitosum (Cav.) DC. (*Crassula magnolii* DC.) - stony slopes - R - Villavallelonga (Anzalone and Bazzichelli 1960), La Brecciosa (Terracciano 1873).
Sedum cepaea L. - cool woods to 800 m - PC
Sedum dasyphyllum L. subsp. ***dasyphyllum*** - walls and cliffs - PC
Sedum dasyphyllum L. subsp. ***glanduliferum*** (Guss.) Nyman - walls and cliffs - PC
Sedum hispanicum L. - walls and cliffs - C
E ***Sedum magellense*** Ten. subsp. ***magellense*** - cool stony slopes - C
Sedum montanum Songeon & E. P. Perrier (*S. ochroleucum* Chaix subsp. *montanum* (Songeon & E. P. Perrier) D. A. Webb) - stony areas and cliffs - R - between Monte Mezzana and the Rifugio! (Conti and Tinti 2012), Valle Ura!, Valle Fredda! Species confirmed for Molise region.
Sedum rubens L. - walls, screes - RR - Vallone Capo d'Acqua! (Campoli Appennino) (Conti 1995).
Sedum rupestre L. - stony areas and cliffs, pastures - CC
Sedum sexangulare L. (*S. boloniense* Loisel.) - walls and cliffs - C
Sempervivum arachnoideum L. (*S. arachnoideum* L. subsp. *tomentosum* (C. B. Lehm. & Schnittsp.) Schinz & Thell.) - cliffs, mountain tops - CC
E ***Sempervivum riccii*** Iberite & Anzal. (*S. italicum* I. Ricci) (Fig. 11) - cliffs, stony slopes - C
Sempervivum tectorum L. - cliffs, stony slopes - C
Umbilicus horizontalis (Guss.) DC. (*Cotyledon horizontalis* Guss.) - humid cliffs, shady walls - PC

Fig. 11 *Sempervivum riccii* (Photo by F. Conti)

Cucurbitaceae

Bryonia dioica Jacq. (*B. cretica* L. subsp. *dioica* (Jacq.) Tutin) - hedges and humid scrub - C

Ecballium elaterium (L.) A. Rich. (*Momordica elaterium* L.) - sandy uncultivated land - PC

Cupressaceae

Juniperus communis L. var. ***communis*** - pastures and open woods, mostly in the middle mountains - CC

Juniperus communis L. var. ***saxatilis*** Pall. (*J. alpina* Gray; *J. communis* L. subsp. *hemisphaerica* (J. Presl & C. Presl) Nyman; *J. communis* L. subsp. *nana* (Willd.) Syme; *J. hemisphaerica* J. Presl & C. Presl; *J. nana* Willd.) - scrub in the subalpine belt - CC

Juniperus deltoides R. P. Adams (*J. oxycedrus* L. subsp. *deltoides* (R. P. Adams) N. G. Passal.; *J. oxycedrus* auct. Fl. Ital. p.p.) - maquis and garigue in Mediterranean and supramediterranean environments - C

Juniperus sabina L. - montane stony slopes - R - Barrea at Colle S. Angelo, near Villavallelonga!, Monte Boccanera, Mainarde on Morrone delle Rose! (Anzalone and Bazzichelli 1960; D'Andrea 1982; Conti et al. 1987; Rovelli 1992).

Cyperaceae

Blysmus compressus (L.) Panz. ex Link (*Schoenus compressus* L.) - humid meadows - C

Carex acuta L. (*C. gracilis* Curtis) - marshy environments - C

Carex acutiformis Ehrh. (*C. paludosa* Gooden.) - marshy environments - RR - riversides of Lacerno (Campoli) (Falqui 1899), Pantanello! (Conti and Minutillo 1998).

Carex brachystachys Schrank - wet cliffs - RR - Camosciara! (Tammaro 1998; Conti 1998).

Carex caryophyllea Latourr. - glades and stony pastures - C

Carex cuprina (Heuff.) A. Kern. (*C. otrubae* Podp.) - ditches, riversides, humid meadows - C

Carex depressa Link subsp. *basilaris* (Jord.) Cif. and Giacom. (*C. basilaris* Jord.) - NC - Gioia Vecchio, Pescasseroli (Anzalone and Bazzichelli 1960).

Carex digitata L. - open woods, glades - PC

Carex distachya Desf. - *Quercus ilex* woods, maquis - R - near Campoli Appennino (Conti 1995), Filignano in loc. Le Mura!, Valle del Rio Chiaro!

Carex distans L. - humid meadows - C

Carex divisa Huds. - marshy environments, humid meadows - RR - La Brionna (Tammaro 1988).

Carex divulsa Stokes subsp. ***divulsa*** - cool woods - PC

Carex elata All. subsp. ***elata*** - marshy environments - R - Piana di Pescasseroli!, Templo! (Venanzoni 1987; Conti 1995), Valle Chiara!

Carex ericetorum Pollich (*C. approximata* All.; *C. ericetorum* Pollich var. *approximata* (All.) Nyman) - NC - Monte Meta, Monte Greco, Monte Forcellone (Tenore and Gussone 1842).

Carex flacca Schreb. - humid meadows - C - subsp. *flacca* and subsp. *serrulata* are both recorded, but in our opinion are not clearly segregated.

Carex flava L. - marshy environments, peaty meadows - R - Lagozzo! (Conti 1994), La Brionna (Tammaro 1988).

Carex frigida All. - sources, stony slopes with running waters - R - Monte Le Rose (Tenore 1842), surroundings of Pescasseroli (Tammaro 1988).

Carex halleriana Asso subsp. ***halleriana*** - thermophilous open woods - PC

Carex hirta L. - humid meadows - PC

Carex humilis Leyss. - stony pastures - PC

Carex kitaibeliana Degen ex Bech. - stony meadows at high altitude, stony slopes - C

Fig. 12 *Carex tomentosa* (Photo by F. Conti)

Carex lasiocarpa Ehrh. - NC - generically recorded for the Park (Sipari 1926 quotes Pirotta; Lusina 1954).

Carex leporina L. (*Carex ovalis* Gooden.) - humid environments - C

Carex liparocarpos Gaudin subsp. ***liparocarpos*** - pastures - R - between Sperone and Colle Biferno!, slopes of Monte Rosa Pinnola! (Conti 1995, 1998).

Carex macrolepis DC. - stony slopes - CC

Carex mucronata All. - cliffs - R - Villavallelonga, Camosciara (Lusina 1954; Tammaro 1988).

Carex muricata L. subsp. ***cesanensis*** A. Mol., Acedo and Llamas - open woods, scrub - PC

Carex muricata L. subsp. ***muricata*** - scrub - RR - Monte Forcellone - Serra Porcarella! (Conti 1992).

Carex nigra (L.) Reichard subsp. ***nigra*** (*C. juncella* (Fr.) Th. Fr.; *C. fusca* All.) - marshy environments - R - Fonte Fredda!, Le Fontane! (Conti 1992).

Carex olbiensis Jord. - humid woods - RR - Bosco Frascaro on the right bank of Rio Chiaro! (Conti and Minutillo 2001).

Carex ornithopoda Willd. - meadows at high altitude - PC

Carex pairae F. W. Schultz (*C. muricata* L. subsp. *lamprocarpa* Čelak.) - D - Samples stored in APP determined as *C. pairae* were recently reviewed by

A. Molina and referred to other taxa belonging to the group of *C. spicata* or *C. muricata*. Consequently, all the bibliographic references of this taxon should be verified.

Carex pallescens L. - humid environments - PC

Carex panicea L. - humid meadows - PC

Carex paniculata L. subsp. ***paniculata*** - marshy environments - PC

Carex pendula Huds. - riparian woods - C

Carex pilosa Scop. - cool woods - RR - Serra Lunga (Scoppola and Modena 1997).

Carex punctata Gaudin - peat - bogs - RR - Valle di Canneto (Terracciano 1878), La Brionna (Tammaro 1988).

Carex remota L. - humid woods - PC

Carex riparia Curtis - watercourses, marshy environments - R - La Brionna (Tammaro 1984a), Cartiera sulle rive del Volturno! (Conti 1992), presso Bisegna (Buchwald 1995).

Carex rostrata Stokes - marshy environments - RR - La Brionna (Tammaro 1984a).

Carex spicata Huds. (*C. contigua* Hoppe) - open woods, margins of woods - PC

Carex sylvatica Huds. subsp. ***sylvatica*** - cool woods - PC

Carex tomentosa L. (*C. filiformis* L.) (Fig. 12) - humid meadows - RR - La Brionna (Tammaro 1988). The nomenclature according to Koopman et al. (2014).

Carex umbrosa Host subsp. ***umbrosa*** (*C. polyrrhiza* Wallr.) - woods - RR - Picinisco a Canneto (Terracciano 1873; Petriglia 2004).

Carex vesicaria L. - marshy environments - R - Il Lagozzo! (Conti 1994), Lago Pantaniello! (Conti 1995).

Carex viridula Michx. (*C. oederi* Retz.; *C. serotina* Mérat) - sources, marshy meadows - PC

Cyperus fuscus L. - humid sands, beds of the rivers - R - “Valvori nell’andare a S. Biagio” (Terracciano 1873), Valleporcina (Picinisco)! (Conti 1995).

Cyperus longus L. - humid environments - RR - near Castel S. Vincenzo (Buchwald 1995).

Eleocharis palustris (L.) Roem. & Schult. subsp. ***palustris*** - marshy environments - C

Eleocharis quinqueflora (Hartmann) O. Schwarz (*S. quinqueflorus* Hartmann) - peaty meadows - RR - Campitelli! (Conti 1994).

Eleocharis uniglumis (Link) Schult. subsp. ***uniglumis*** (*Scirpus uniglumis* Link) - marshy environments - R - Campitelli!, between Pescasseroli and Bisegna in loc. Fontane della Padura! (Conti 1998; Conti and Minutillo 1998).

Eriophorum latifolium Hoppe (Fig. 13) - peaty meadows - R - Val Fondillo!, Camosciara! (Lusina 1954; Anzalone and Bazzichelli 1960; Conti 1998).

Isolepis cernua (Vahl) Roem. & Schult. (*Scirpus cernuus* Vahl; *S. savii* Sebast. & Mauri) - humid sands - R - Fonte Scopella! (Conti and Minutillo 2001), Monte Marsicano (Pignotti 2003).

Pycreus flavescens (L.) P. Beauv. ex Rchb. (*Cyperus flavescens* L.) - observed at Rio Mollarino! (Conti and Minutillo 1998) outside the Park but close to the buffer external zone.

Fig. 13 *Eriophorum latifolium* (Photo by F. Conti)

Schoenoplectus lacustris (L.) Palla subsp. ***glaucus*** (Sm.) Luceño & Marín (*Scirpus lacustris* L. subsp. *tabernaemontani* (C. C. Gmel.) Syme; *Scirpus glaucus* Sm.; *Schoenoplectus tabernaemontani* (C. C. Gmel.) Palla; *Scirpus tabernaemontani* C. C. Gmel.) - marshy environments - PC

Schoenoplectus lacustris (L.) Palla subsp. ***lacustris*** - marshy environments - PC

Schoenus nigricans L. - humid meadows - RR - Lago Cardito (Conti and Minutillo 1998).

Scirpoides holoschoenus (L.) Soják (*Holoschoenus australis* (L.) Rchb.; *Scirpoides holoschoenus* subsp. *australis* (Murray) Soják; *Holoschoenus romanus* (L.) Fritsch; *Scirpus holoschoenus* L.; *Scirpus australis* L.; *Scirpus romanus* L.) - marshy environments - C

D

Dennstaedtiaceae

Pteridium aquilinum (L.) Kuhn subsp. ***aquilinum*** (*Pteris aquilina* L.) - clearings and abandoned cultivated land - CC

Dioscoreaceae

Dioscorea communis (L.) Caddick & Wilkin (*Tamus communis* L.) - woods - C

Dryopteridaceae

Dryopteris filix-mas (L.) Schott (*Polypodium filix-mas* L.; *Aspidium filix-mas* (L.) Sw.; *Nephrodium filix-mas* (L.) Rich.; *Lastrea filix-mas* (L.) C. Presl) - woods - C

Dryopteris pallida (Bory) Maire & Petitm. subsp. ***pallida*** (*D. villarii* (Bellardi) Woynar subsp. *pallida* (Bory) Heywood; *Nephrodium pallidum* Bory) - shady cliffs, usually to 1,000 m - PC

Dryopteris submontana(Fraser-Jenk. & Jermy) Fraser-Jenk. (*D. villarii* Bellardi subsp. *submontana*Fraser-Jenk. & Jermy) - shady stony slopes - RR - Cava di Rena! (Conti and Soldati 2010).

Dryopteris villarii (Bellardi) Woyn. ex Schinz & Thell. subsp. ***villarii*** (*Polypodium villarii* Bellardi) - screes, cool montane stony slopes - PC

Polystichum aculeatum (L.) Roth (*Polypodium aculeatum* L.; *Dryopteris lobata* (Huds.) Schinz & Thell.; *D. aculeata* Kuntze subsp. *lobata* (Huds.) Briq.; *Aspidium aculeatum* (L.) Sw.) - woods - C

Polystichum lonchitis (L.) Roth (*Polypodium lonchitis* L.) - cool *Fagus sylvatica* woods and shady cliffs, usually over 1,500 m - C

Polystichum setiferum (Forssk.) Woyn. (*Polypodium setiferum* Forssk.) - woods - CC

F. Conti, F. Bartolucci, *The Vascular Flora of the National Park of Abruzzo, Lazio and Molise (Central Italy)*, Geobotany Studies, DOI 10.1007/978-3-319-09701-5_10

E

Ephedraceae

Ephedra major Host subsp. ***major*** (*E. nebrodensis* Tineo ex Guss.) (Fig. 1) - cliffs and stony slopes - R - *ex Praetutiis in montibus Marsorum* (Bertoloni 1833–1854), Gole del Sagittario! (Anzalone 1961), Ortona dei Marsi! (Conti 1995).

Equisetaceae

Equisetum arvense L. - humid environments - C
Equisetum fluviatile L. - humid meadows, rivers - PC
Equisetum hyemale L. - humid woods and meadows - RR - Camosciara! (Lusina 1954; Anzalone and Bazzichelli 1960)
Equisetum palustre L. - humid meadows and marshy environments - C
Equisetum ramosissimum Desf. subsp. ***ramosissimum*** - humid uncultivated land, ditches - PC
Equisetum telmateia Ehrh. - humid environments - CC

Ericaceae

Arbutus unedo L. - maquis and *Quercus ilex* woodland - R - near the Lago di Grotta Campanaro (Spada 1979), Monte Falconara!
Arctostaphylos uva-ursi (L.) Spreng. (*Arbutus uva-ursi* L.) - subalpine scrub, stony slopes - PC
Erica multiflora L.subsp. ***multiflora*** - maquis and garigue - R - "Valvori nell'andare a S. Biagio Saracinisco" (Terracciano 1873 sub *E. ramulosa* Viv.), Lago Grotta Campanaro (Spada 1979), Colleruta!

F. Conti, F. Bartolucci, *The Vascular Flora of the National Park of Abruzzo, Lazio and Molise (Central Italy)*, Geobotany Studies, DOI 10.1007/978-3-319-09701-5_11

Fig. 1 *Ephedra major* subsp. *major* (Photo by F. Conti)

Moneses uniflora (L.) A. Gray (*Pyrola uniflora* L.) - scrub with *Pinus mugo* - R - Camosciara!, M. Meta (Anzalone and Bazzichelli 1960; Bruno and Bazzichelli 1966; Stanisci 1994).

****Monotropa hypophegea*** Wallr. - *Fagus sylvatica* woods, cool woods - R - Valle Inguagnera!, loc. Aia Santilli, presso la Camosciara!

Monotropa hypopitys L. - *Fagus sylvatica* woods, cool woods - PC

Orthilia secunda (L.) House (*Pyrola secunda* L.) - *Fagus sylvatica* woods - C

Pyrola chlorantha Sw. - *Fagus sylvatica* woods - RR - Camosciara! (Conti 1998).

Pyrola minor L. - cool *Fagus sylvatica* woods - PC

Vaccinium myrtillus L. - *Fagus sylvatica* woods in the higher belt - PC

Euphorbiaceae

Euphorbia amygdaloides L. subsp. ***amygdaloides*** - cool woods, mainly *Fagus sylvatica* woods - CC

Euphorbia characias L. - thermophilous stony slopes, maquis, garigue - R - Monte S. Marcello (Falqui 1899), Monte Mattone over Pizzone! (Conti 1995).

E ***Euphorbia corallioides*** L. - woods - PC

Euphorbia cyparissias L. - stony slopes, pebbly areas, arid pastures - CC

Euphorbia dulcis L. (incl. *E. dulcis* subsp. *incompta* (Ces.) Nyman; incl. *E. dulcis* subsp. *purpurata* (Thuill.) Rothm.; incl. *E. purpurata* Thuill.; *Tithymalus dulcis* (L.) Scop.) - cool woods - PC

Euphorbia exigua L. subsp. ***exigua*** (*Tithymalus exiguus* (L.) Hill) - arid meadows, ruderal environments, garigue, fields - PC

Euphorbia falcata L. subsp. ***falcata*** (*Tithymalus falcatus* (L.) Klotzsch and Garcke) - fields, arid uncultivated land - PC

E ***Euphorbia gasparrinii*** Boiss. subsp. ***samnitica*** (Fiori) Pignatti (*E. epithymoides* L. var. *samnitica* Fiori) (Fig. 7 in chapter "Vegetation Features") - cool pastures - PC

Euphorbia helioscopia L. subsp. ***helioscopia*** - fields, ruderal environments - C

Euphorbia myrsinites L. subsp. ***myrsinites*** - stony slopes - C

Euphorbia nicaeensis All. subsp. ***nicaeensis*** (*E. nicaeensis* All. var. *prostrata* Fiori; *E. nicaeensis* All. subsp. *prostrata* (Fiori) Arrigoni) - stony slopes - PC

Euphorbia peplus L. (incl. *E. peploides* Gouan; *Tithymalus peplus* (L.) Hill) - fields, ruderal environments - PC

Euphorbia platyphyllos L. (incl. *E. literata* Jacq.; incl. *E. platyphyllos* L. subsp. *literata* (Jacq.) Holub) - uncultivated land, fields - C

Euphorbia spinosa L. - sunny stony slopes - PC

Mercurialis annua L. - ruderal environments - C

Mercurialis ovata Sternb. and Hoppe - screes and stony pastures - PC

Mercurialis perennis L. - cool woods - C

F

Fabaceae

A ***Amorpha fruticosa*** L. - watercourses - NAT - riversides of Volturno river, near Castel S. Vincenzo! (Conti 1992).

Anthyllis montana L. subsp. ***jacquinii*** (A. Kern.) Hayek (*A. jacquinii* A. Kern.; *A. montana* L. var. *atropurpurea* Vuk.; *A. montana* L. subsp. *atropurpurea* (Vuk.) Pignatti) - stony slopes at high altitude - C

Anthyllis vulneraria L. subsp. ***carpatica*** (Pant.) Nyman (*A. carpatica* Pant.) - montane pastures - PC

Anthyllis vulneraria L. subsp. ***polyphylla*** (DC.) Nyman (*A. vulneraria* L. var. *polyphylla* DC.) - stony montane pastures - PC

Anthyllis vulneraria L. subsp. ***pulchella*** (Vis.) Bornm. (*A. vulneraria* L. var. *pulchella* Vis.) - meadows at high altitude - PC

Anthyllis vulneraria L. subsp. ***rubriflora*** (DC.) Arcang. (*A. vulneraria* L. subsp. *praepropera* (A. Kern.) Bornm.; *A. vulneraria* L. var. *rubriflora* DC.) - montane pastures - C

Anthyllis vulneraria L. subsp. ***weldeniana*** (Rchb.) Cullen (*A. weldeniana* Rchb.) - stony pastures - C

Argyrolobium zanonii (Turra) P. W. Ball subsp. ***zanonii*** (*Cytisus zanonii* Turra) - arid meadows, garigue - PC

*E ***Astragalus aquilanus*** Anzal. (Fig. 1) - stony slopes, margins of termophilous woods, roadsides - RR - between Gioia dei Marsi and Gioia Vecchio!

Astragalus danicus Retz. (Fig. 2) - humid meadows - RR - Piano Aremogna near Lago di Castello! (Conti 1995).

Astragalus depressus L. subsp. ***depressus*** - montane pastures - C

Astragalus glycyphyllos L. - deciduous woods, uncultivated land - C

Astragalus hamosus L. - arid pastures and meadows - PC

F. Conti, F. Bartolucci, *The Vascular Flora of the National Park of Abruzzo, Lazio and Molise (Central Italy)*, Geobotany Studies, DOI 10.1007/978-3-319-09701-5_12

Fig. 1 *Astragalus aquilanus* (Photo by F. Conti)

Fig. 2 *Astragalus danicus* (Photo by F. Conti)

Astragalus hypoglottis L. subsp. *gremlii* (Burnat) Greuter and Burdet - D - surroundings of Gioia Vecchio, Valico di Pantano (Vaccari and Wilczek 1940).

Astragalus monspessulanus L. subsp. ***monspessulanus*** - arid uncultivated land - PC

Astragalus muelleri Steud. & Hochst. - Cocullo (Greco and Petriccione 1989). To be excluded.

Astragalus sempervirens Lam. (incl. *A. sempervirens* Lam. subsp. *gussonei* Pignatti) - stony montane pastures - C

Astragalus sesameus L. - arid meadows and garigue - PC

E ***Astragalus sirinicus*** Ten. - stony pastures at high altitude - RR - Passo dei Monaci! (Tenore 1835–1838; Tenore and Gussone 1842).

Astragalus vesicarius L. subsp. ***vesicarius*** - meadows at high altitude, montane pastures C

A ***Cercis siliquastrum*** L. subsp. ***siliquastrum*** - roadsides, stony slopes, thermophilous oak - woods to 800 m - NAT

*A ***Cicer arietinum*** L. - fields - CAS - Villavallelonga, between the village and Madonna della Lanna, loc. Cona Rovara!

Colutea arborescens L. (*C. brevialata* Lange) - open oak - woods and hedges to 1,200 m - PC

Coronilla minima L. subsp. ***minima*** - arid meadows - PC

Coronilla repanda (Poir.) Guss. subsp. ***repanda*** (*Ornithopus repandus* Poir.) - arid uncultivated land - RR - Villetta Barrea (Anzalone and Bazzichelli 1960).

Coronilla scorpioides (L.) W. D. J. Koch (*Ornithopus scorpioides* L.) - arid meadows - C

Coronilla vaginalis Lam. - stony slopes - PC

Cytisophyllum sessilifolium (L.) O. Lang (*Cytisus sessilifolius* L.) - open woods, margins of woods and scrub - C

Cytisus decumbens (Durande) Spach (*Spartium decumbens* Durande) - stony pastures and open woods on the rocks - PC

Cytisus hirsutus L. (*Chamaecytisus hirsutus* (L.) Link; incl. *Ch. polytrichus* (M. Bieb.) Rothm.; *Ch. triflorus* (Lam.) Skalická) - arid pastures, maquis and scrub - PC

Cytisus spinescens C. Presl (*Chamaecytisus spinescens* (C. Presl) Rothm.) - stony slopes, garigue, maquis and open woods - C

Cytisus villosus Pourr. - maquis and hill and montane woods in shady and cool areas - R - La Giostra, Monte Annamunna, Monte Ara dei Merli (Petriccione et al. 1994), Valle di Amplero and Vallelonga (Tammaro 1998).

Emerus major Mill. subsp. ***emeroides*** (Boiss. & Spruner) Soldano & F. Conti (*Coronilla emeroides* Boiss. & Spruner; *C. emerus* L. subsp. *emeroides* (Boiss. & Spruner) Hayek; *Hippocrepis emerus* (L.) Lassen subsp. *emeroides* (Boiss. & Spruner) Lassen) - open woods and margins of xerophilous woods - PC

Emerus major Mill. subsp. ***major*** (*Coronilla emerus* L. subsp. *emerus*; *Hippocrepis emerus* (L.) Lassen subsp. *emerus*) - open woods and margins of mesophilous woods - C

Galega officinalis L. - humid uncultivated land - PC

Genista januensis Viv. subsp. *januensis* - NC - alla Difensa (Grande 1904).

Genista radiata (L.) Scop. (*Spartium radiatum* L.) - NC - Villavallelonga (Grande 1910).

Genista sagittalis L. (*Chamaespartium sagittale* (L.) Gibbs) - montane pastures - R - Monte Genzana! (Conti 1998), Bocche Chiarano, Serra Santa Maria, Serra del Feudo (Di Pietro et al. 2005).

Genista tinctoria L. (incl. *G. ovata* Waldst. & Kit.) - open woods and their borders - C
Hippocrepis biflora Spreng. (*H. unisiliquosa* L.) - arid meadows - R - Picinisco (Tenore and Gussone 1842), near Rocchetta al Volturno, near theAbbazia of Castel S. Vincenzo! (Conti 1995).
Hippocrepis comosa L. subsp. ***comosa*** - arid meadows - CC
Hippocrepis glauca Ten. - arid meadows - PC
Laburnum alpinum (Mill.) Bercht. & J. Presl (*Cytisus alpinus* Mill.) - *Fagus sylvatica* woods glades - PC
Laburnum anagyroides Medik. subsp. ***anagyroides*** - margins of woods and glades from the supramediterranean to the mountain belt - CC
Lathyrus annuus L. - uncultivated land, fields - PC
Lathyrus aphaca L. subsp. ***aphaca*** - uncultivated land - C
Lathyrus cicera L. - arid uncultivated land, fields - R - Gioia Vecchio (Anzalone and Bazzichelli 1960), Frattura! (from a specimen collected by Paolessi and keeped in APP), bassa Valle Macrana!
Lathyrus hirsutus L. - uncultivated land, pastures - R - Villavallelonga! (Grande 1904), Templo!
Lathyrus inconspicuus L. - arid uncultivated land - R - Villavallelonga (Grande 1904), Gole del Sagittario! (Conti and Tinti 2012).
Lathyrus latifolius L. - uncultivated land, hedges - PC
Lathyrus niger (L.) Bernh. (*Orobus niger* L.) - cool woods on slightly acid soil - R - from Pizzone to Alfedena, near the way to Montenero (Paura and Abbate 1995), Montaquila at Bosco Frascaro! (Conti and Minutillo 2001).
Lathyrus nissolia L. - uncultivated land, humid meadows - PC
Lathyrus ochrus (L.) DC. (*Pisum ochrus* L.) - NC - Picinisco, Barrea (Tenore and Gussone 1842).
Lathyrus pannonicus (Jacq.) Garcke (*Orobus pannonicus* Jacq.) (Fig. 3) - humid meadows - R - between Pescasseroli and Bisegna! (Conti 1992), Villavallelonga (Pignatti 1982). According Schlee et al. (2011) is currently not possible to attribute the C - Apennine populations to subspecies known.
Lathyrus pratensis L. subsp. ***pratensis*** - meadows, scrub - C
Lathyrus setifolius L. - arid uncultivated land - PC
Lathyrus sphaericus Retz. - arid uncultivated land - PC
Lathyrus sylvestris L. subsp. ***sylvestris*** - hedges, margins of woods - PC
Lathyrus venetus (Mill.) Wohlf. (*Orobus venetus* Mill.) - woods - C
Lathyrus vernus (L.) Bernh. subsp. ***vernus*** - woods - C
Lens ervoides (Brign.) Grande (*Cicer ervoides* Brign.) - arid meadows - PC
Lens nigricans (M. Bieb.) Godr. (*Ervum nigricans* M. Bieb.) - arid meadows - R - Fosso Macrana!, Gioia Vecchio!, (Conti and Minutillo 1998), ridge over Castrovalva!
Lotus corniculatus L. subsp. ***alpinus*** (DC.) Rothm. (*L. alpinus* (DC.) Schleicher; *L. corniculatus* L. var. *alpinus* DC.) - alpine pastures - PC
Lotus corniculatus L. subsp. ***corniculatus*** - arid pastures, uncultivated land - C

Fig. 3 *Lathyrus pannonicus* (Photo by F. Conti)

Lotus herbaceus (Vill.) Jauzein (*Dorycnium pentaphyllum* Scop. subsp. *herbaceum* (Vill.) Rouy; *D. herbaceum* Vill.; *D. pentaphyllum* Scop. subsp. *herbaceum* (Vill.) Rouy) - arid meadows - PC

Lotus pedunculatus Cav. (*L. uliginosus* Schkuhr) - NC - surroundings of Gioia Vecchio (Vaccari and Wilczek 1940).

Lotus tenuis Waldst. & Kit. ex Willd. (*L. glaber* Mill.) - humid meadows - PC

A ***Lupinus albus*** L. subsp. ***albus*** - CAS

Medicago arabica (L.) Huds. (*M. polymorfa* L. var. *arabica* L.) - uncultivated land, fields - PC

Medicago disciformis DC. - garigue - RR - Valle del Sagittario, Castrovalva - Colle S. Michele! (Conti and Tinti 2012).

Medicago falcata L. subsp. ***falcata*** (*M. sativa* L. subsp. *falcata* (L.) Arcang.) - uncultivated land, roadsides - PC

Medicago lupulina L. (incl. *M. lupulina* L. subsp. *cupaniana* (Guss.) Nyman; incl. *M. cupaniana* Guss.) - pastures, uncultivated land, ruderal environments - C

Medicago minima (L.) L. (*M. polymorpha* L. var. *minima* L.) - arid meadows - C

Medicago murex Willd. (*M. sphaerocarpos* Bertol.) - pastures - RR - Campoli Appennino in loc. Querceto! (Conti and Minutillo 1998).

Medicago orbicularis (L.) Bartal. (*M. polymorpha* L. var. *orbicularis* L.) - fields, arid uncultivated land - PC
Medicago polymorpha L. (*M. hispida* Gaertn.) - uncultivated land, fields - PC
Medicago prostrata Jacq. subsp. ***prostrata*** - arid and stony meadows - PC
Medicago rigidula (L.) All. (*M. polymorfa* L. var. *rigidula* L.) - arid meadows - R - Monte Mattone near Pizzone! (Conti 1995), near the Abbazia of Castel S. Vincenzo!, between Collelongo and Trasacco!
Medicago sativa L. - uncultivated land, roadsides - PC
Melilotus albus Medik. - uncultivated land - C
Melilotus altissimus Thuill. - humid uncultivated land - C
Melilotus indicus (L.) All. (*Trifolium indicum* L.) - fields, uncultivated land - R - Cocullo (Greco and Petriccione 1989), Fossato di Rosa (Petriccione et al. 1994).
Melilotus neapolitanus Ten. - sandy arid uncultivated land - PC
Melilotus officinalis (L.) Lam. (*Trifolium officinale* L.) - uncultivated land - C
Melilotus sulcatus Desf. - arid uncultivated land - PC
Onobrychis alba (Waldst. & Kit.) Desv. subsp. ***alba*** (*O. alba* (Waldst. & Kit.) Desv. subsp. *tenoreana* (Lacaita) Pignatti; *O. tenoreana* Lacaita) - arid pastures - C
Onobrychis caput-galli (L.) Lam. - arid uncultivated land - R - Colleruta (Conti and Minutillo 1998), Monte S. Angelo near Pizzone!
Onobrychis viciifolia Scop. - ruderal environments, roadsides - C
Ononis natrix L. subsp. *natrix* - NC - surroundings of Gioia Vecchio, between Gioia Vecchio and Pescasseroli (Vaccari and Wilczek 1940).
Ononis pusilla L. subsp. ***pusilla*** - arid meadows - C
Ononis reclinata L. (incl. *O. reclinata* L. subsp. *mollis* (Savi) Bég.) - arid meadows - PC
Ononis spinosa L. subsp. ***spinosa*** - arid clayey uncultivated land - PC
Ononis viscosa L. subsp. ***breviflora*** (DC.) Nyman (*O. breviflora* DC.; *O. viscosa* L. subsp. *brachycarpa* (DC.) Batt.) - arid pastures - PC
Oxytropis campestris (L.) DC. (*Astragalus campestris* L.) - stony meadows at high altitude - PC
E ***Oxytropis pilosa*** (L.) DC. subsp. ***caputoi*** (Moraldo & La Valva) Brilli - Catt., Di Massimo & Gubellini (*O. caputoi* Moraldo & La Valva) (Fig. 4) - stony meadows at high altitude - PC
Pisum sativum L. subsp. ***biflorum*** (Raf.) Soldano (*P. biflorum* Raf.; *P. elatius* M. Bieb.; *P. sativum* L. subsp. *elatius* (M. Bieb.) Asch. & Graebn.) - uncultivated land - PC
A ***Robinia pseudoacacia*** L. - roadsides - INV
Scorpiurus subvillosus L. - arid uncultivated land - PC
Securigera securidaca (L.) Degen and Dörfl. (*Coronilla securidaca* L.) - arid uncultivated land - PC
Securigera varia (L.) Lassen (*Coronilla varia* L.) - open woods, glades - PC
Spartium junceum L. - scrub in the coast and hill belt - CC
Sulla coronaria (L.) Medik. (*Hedysarum coronarium* L.) - uncultivated land - PC
Trifolium alpestre L. - glades, open woods - PC

Fig. 4 *Oxytropis pilosa* subsp. *caputoi* (Photo by F. Conti)

Trifolium angustifolium L. subsp. ***angustifolium*** - pastures and arid uncultivated areas - PC

Trifolium arvense L. subsp. ***arvense*** - arid uncultivated land - C

Trifolium aureum Pollich subsp. ***aureum*** - wood glades - PC

*****Trifolium bocconei*** Savi - arid uncultivated land - RR - Capo d'Acqua near Campoli Appennino!

Trifolium campestre Schreb. - meadows and arid uncultivated land - C

Trifolium dubium Sibth. (*T. minus* Sm.) - humid meadows - R - Colle dell'Acquaro near Picinisco (Tenore 1842 sub *T. minus* Sm.), La Brionna (Tammaro 1984), Piazzale Canneto!

Trifolium fragiferum L. subsp. ***fragiferum*** - humid meadows - PC

Trifolium glomeratum L. - meadows - RR - Campoli Appennino in loc. Querceto! (Conti and Minutillo 1998).

Trifolium hybridum L. subsp. ***elegans*** (Savi) Asch. & Graebn. (*T. elegans* Savi) - uncultivated land - RR - Gole del Sagittario near the old mill! (Conti and Tinti 2012).

Trifolium hybridum L. subsp. ***hybridum*** - humid meadows - R - from Pescasseroli to Bisegna! (Conti 1994, 1998), Vallecupa!

A ***Trifolium incarnatum*** L. subsp. ***incarnatum*** - uncultivated land - NAT

Trifolium incarnatum L. subsp. ***molinerii*** (Hornem.) Ces. (*T. molinerii* Hornem.) - uncultivated land - PC

Trifolium lucanicum Gasp. (*T. scabrum* L. subsp. *lucanicum* (Gasp.) Pignatti) - arid meadows - PC

Trifolium medium L. subsp. ***medium*** - meadows and margins of woods - PC

Trifolium micranthum Viv. (*T. filiforme* L.) - humid pastures - R - Piazzale Canneto (Conti 1995), Montenero Val Cocchiara!

Trifolium montanum L. subsp. ***rupestre*** (Ten.) Nyman (*T. rupestre* Ten.) - stony pastures - C

Trifolium nigrescens Viv. subsp. ***nigrescens*** - arid pastures - PC

Trifolium ochroleucon Huds. - meadows and glades - C

Trifolium patens Schreb. - humid meadows - R - Picinisco (Tenore and Gussone 1842; Grande 1925), Pantanello! (Conti and Minutillo 1998), Scontrone (Griebl 2010).

Trifolium phleoides Willd. - arid pastures - R - Villavallelonga (Grande 1914), Le Forme! (Conti 1995).

Trifolium pratense L. subsp. ***pratense*** - meadows, uncultivated land - C

E ***Trifolium pratense*** L. subsp. ***semipurpureum*** (Strobl) Pignatti (*T. pratense* L. var. *semipurpureum* Strobl) - subalpine pastures - CC

Trifolium repens L. (incl. *T. repens* L. subsp. *prostratum* Nyman) - uncultivated land, cool meadows - CC

Trifolium resupinatum L. (incl. *T. suaveolens* Willd.) - humid meadows - C

Trifolium rubens L. - *Fagus sylvatica* wood glades - R - from Picinisco to the Santuario di Canneto! (Conti and Minutillo 1998).

Trifolium scabrum L. subsp. ***scabrum*** - arid meadows - PC

Trifolium squarrosum L. (incl. *T. marsicum* Ten.) - arid pastures - R - arid pastures of Marsica (Tammaro and Pace 1994).

Trifolium stellatum L. - arid uncultivated land, ruderal environments - PC

****Trifolium striatum*** L. subsp. ***striatum*** - uncultivated land, pastures - RR - Valle Canale near Collelongo!

Trifolium strictum L. - NC - Villavallelonga (Aia dei Merli) (Anzalone and Bazzichelli 1960 from a specimen collected by Grande).

Trifolium subterraneum L. (incl. *T. subterraneum* L. subsp. *yanninicum* Katzn. & F. H. W. Morley) - arid uncultivated land - R - between Lagone and Mastrogiovanni!, Montenero Val Cocchiara! (Conti and Minutillo 2001).

Trifolium thalii Vill. - stony slopes, damp snowbed meadows - C

A *Trigonella foenum-graecum* L. - uncultivated land - CAS - Colle delle Marie (Viegi et al. 1990 from a specimen collected by Grande).

Trigonella gladiata M. Bieb. (*T. gladiata* Steven) - arid meadows - PC

Trigonella monspeliaca L. (*Medicago monspeliaca* (L.) Trautv.) - arid meadows - R - near Pizzone! (Conti 1992), between Collelongo and Trasacco!, near Pescina!

Vicia bithynica (L.) L. - uncultivated land, arid meadows, fields - PC

Vicia cassubica L. - woods - R - Rio Freddo (Tenore 1830), generically recorded for the Park (Anzalone and Bazzichelli 1960).

Vicia cracca L. - open woods, glades - PC
Vicia disperma DC. - uncultivated areas, arid pastures - R - Pescasseroli (Anzalone and Bazzichelli 1960), Monte della Rocchetta!
A ***Vicia ervilia*** (L.) Willd. (*Ervum ervilia* L.) - cultivated land - NAT - Villetta Barrea (Anzalone and Bazzichelli 1960), near Gioia Vecchio (Conti 1995).
*****Vicia grandiflora*** Scop. (*V. grandiflora* Scop. subsp. *sordida* (Waldst. & Kit.) Dostál) - degraded woods, hedges, uncultivated land - R - Valle del Rio Chiaro!, Filignano in loc. Le Mura!
Vicia hirsuta (L.) Gray (*Ervum hirsutum* L.) - arid pastures - PC
Vicia hybrida L. - arid meadows - PC
Vicia incana Gouan (*V. cracca* L. subsp. *incana* (Gouan) Rouy) - open woods, glades - C
Vicia laeta Ces. (*V. barbazitae* Ten. & Guss.) - NC - Monte La Difesa (Fiori 1928 sub *V. grandiflora* Scop. var. *laeta* (Ces.)).
Vicia lathyroides L. - meadows and arid uncultivated land - RR - Gole del Sagittario! (Conti 1998).
Vicia lutea L. (incl. *V. vestita* Boiss.; *V. lutea* L. subsp. *vestita* (Boiss.) Rouy) - uncultivated land, arid pastures - PC
Vicia narbonensis L. subsp. ***narbonensis*** - uncultivated and cultivated land - R - Campoli Appennino!, Monte Falconara (Conti and Minutillo 2001).
Vicia onobrychioides L. - coppices, hedges, pastures - PC
Vicia pannonica Crantz subsp. ***striata*** (M. Bieb.) Nyman (*V. striata* M. Bieb.) - uncultivated land, hedges - C
Vicia peregrina L. - arid uncultivated land, fields - PC
Vicia sativa L. subsp. *cordata* (Hoppe) Batt. (*V. cordata* Hoppe) - NC - Valico di Pantano (Vaccari and Wilczek 1940).
Vicia sativa L. subsp. ***nigra*** (L.) Ehrh. (*V. sativa* subsp. *angustifolia* (Grufb.) Gaudin; *V. sativa* var. *nigra* L.; *V. sativa* subsp. *segetalis* (Thuill.) Gaudin; *V. segetalis* Thuill.) - uncultivated land, arid meadows - C
Vicia sativa L. subsp. ***sativa*** - uncultivated land - RR - Gole del Sagittario! (Conti and Tinti 2012).
Vicia sepium L. - woods, glades - C
Vicia tenuifolia Roth subsp. ***tenuifolia*** - glades - PC
Vicia villosa Roth subsp. ***varia*** (Host) Corb. (*V. varia* Host) - hedges, scrub - C

Fagaceae

Castanea sativa Mill. - acidophil woods of the montane belt - PC
Fagus sylvatica L. subsp. ***sylvatica*** - woods of the montane belt - CC
Quercus cerris L. - woods to 1,200 m - CC
Quercus dalechampii Ten. - heliophilous and thermophilous woods - R - near the Abbazia of Castel S. Vincenzo!, between S. Gennaro and Valle Porcina! (Conti 1995).

Quercus frainetto Ten. - thermo-mesophilous woods - R - Valle Lunga near S. Elia (Terracciano 1873), between Picinisco and Valleporcina! (Spada 1979), Fontitune, Mainarde (Conti 1995).

Quercus ilex L. subsp. ***ilex*** - woods and maquis on thermophilous stony slopes to the low montane belt - PC

Quercus petraea (Matt.) Liebl. subsp. ***petraea*** (*Q. robur* L. var. *petraea* Matt.) - mesophile woods to 1,000 m - RR - Gole del Sagittario! (Conti and Tinti 2012).

Quercus pubescens Willd. subsp. ***pubescens*** (incl. *Q. virgiliana* (Ten.) Ten.) - heliophilous and thermophilous woods to 900–1,000 m (rarely to 1,200–1,500 m) - CC

Quercus robur L. subsp. ***robur*** - plain - growing and sporadically mesophile hill woods - RR - generically recorded for the Park (D'Andrea 1982), Cocullo (Greco and Petriccione 1989).

Quercus x pseudosuber Santi (*Q. crenata* auct.) - margins of woods - RR - S. Sebastiano near Madonna del Calvario! (Conti and Minutillo 2001).

G

Gentianaceae

Blackstonia perfoliata (L.) Huds. subsp. ***perfoliata*** - humid environments, fields - R - Mainarde (Zodda 1931), Villetta Barrea (Anzalone and Bazzichelli 1960), Frattura!

Centaurium erythraea Rafn subsp. ***erythraea*** - fields, pastures, garigue - PC

Centaurium pulchellum (Sw.) Druce subsp. *pulchellum* (*Erythraea ramosissima* (Vill.) Pers.) - NC - Picinisco (Tenore and Gussone 1842). The report for Valleporcina (Conti 1995) is to be referred to *C. tenuiflorum* subsp. *acutiflorum*.

Centaurium tenuiflorum (Hoffmanns. & Link) Fritsch subsp. ***acutiflorum*** (Schott) Zeltner - humid uncultivated land - R - Villavallelonga (Anzalone and Bazzichelli 1960), between Coll'Alto and Scapoli! (Conti 1995), Valleporcina! (Conti 1995 sub *C. pulchellum* subsp. *pulchellum*).

Gentiana brachyphylla Vill. subsp. *favratii* (Rittener) Tutin (*G. magellensis* (Vaccari ex Ronniger) Tammaro; *G. orbicularis* Schur; *G. verna* L. subsp. *brachyphylla* (Vill.) P. Fourn.) - NC - the report of *G. bavarica* for Monte Greco (Tenore and Gussone 1842) is probably to be referred to this taxon.

Gentiana cruciata L. subsp. ***cruciata*** - pastures - C

Gentiana dinarica Beck - stony slopes - C

Gentiana lutea L. subsp. ***lutea*** - montane pastures - C

Gentiana nivalis L. - cool meadows at high altitude - PC

Gentiana utriculosa L. - stony meadows at high altitude - PC

Gentiana verna L. subsp. ***verna*** - cool montane pastures, stony slopes - CC

E ***Gentianella columnae*** (Ten.) Holub (*G. columnae* Ten.) (Fig. 1) - meadows at high altitude - C - the reports of *G. campestris*, *G. anisodonta* and *G. amarella* are to be referred to this taxon.

F. Conti, F. Bartolucci, *The Vascular Flora of the National Park of Abruzzo, Lazio and Molise (Central Italy)*, Geobotany Studies, DOI 10.1007/978-3-319-09701-5_13

Fig. 1 *Gentianella columnae* (Photo by F. Conti)

Gentianopsis ciliata (L.) Ma subsp. ***ciliata*** (*Gentiana ciliata* L. subsp. *ciliata*; *Gentianella ciliata* (L.) Borkh. subsp. *ciliata*) - montane pastures, stony slopes - C

Geraniaceae

Erodium acaule (L.) Bech. & Thell. (*Geranium romanum* Burm. f.; *E. romanum* (Burm. f.) L'Hér.) - uncultivated land, arid pastures - PC

E ***Erodium alpinum*** (Burm. f.) L'Hér. (*Geranium alpinum* Burm. f.) (Fig. 2) - pastures - PC

Erodium ciconium (L.) L'Hér. - arid pastures, ruderal environments - PC

Erodium cicutarium (L.) L'Hér. - arid uncultivated land, ruderal environments - C

Erodium malacoides (L.) L'Hér. subsp. ***malacoides*** - ruderal environments - PC

Erodium moschatum (L.) L'Hér. - NC - Picinisco (Tenore and Gussone 1842).

E ***Geranium austroapenninum*** Aedo (*G. cinereum* auct. Fl. Ital.; *G. subcaulescens* auct. Fl. Ital.) (Fig. 3) - stony meadows - C

Geranium columbinum L. - uncultivated land, ruderal environments, cultivated areas - C

Geranium dissectum L. - uncultivated land - C

Geranium lanuginosum Lam. - NC - Villavallelonga (Aedo et al. 2007 from a specimen collected by Fiori).

Geranium lucidum L. - walls, shady cliffs - C

Geranium macrorrhizum L. - screes, shady cliffs - PC

Geranium molle L. (*G. molle* L. subsp. *brutium* (Gasp.) Graebn.) - ruderal environments, uncultivated land - C

Geranium purpureum Vill. (*G. robertianum* L. subsp. *purpureum* (Vill.) Nyman) - cliffs, hedges, scrub - C

Geranium pusillum L. - ruderal environments - R - between Pescasseroli and Barrea (Vaccari and Wilczek 1940), Villavallelonga (Anzalone and Bazzichelli 1960).

Fig. 2 *Erodium alpinum* (Photo by F. Conti)

Fig. 3 *Geranium austroapenninum* (Photo by F. Conti)

Geranium pyrenaicum Burm. f. subsp. ***pyrenaicum*** - uncultivated land, margins of woods - CC

Geranium reflexum L. (Fig. 4) - clear areas of *Fagus sylvatica* woods with accumulation of organic humus - PC

Geranium robertianum L. - woods, hedges, shady cliffs, walls - C

Geranium rotundifolium L. - cultivated land, hedges - C

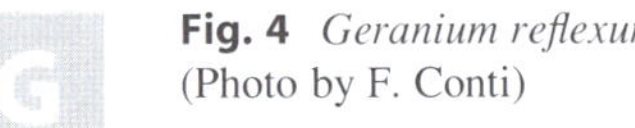

Fig. 4 *Geranium reflexum* (Photo by F. Conti)

Geranium sanguineum L. - scrub, glades - PC
Geranium sylvaticum L. - damp grassland, *Fagus sylvatica* wood glades - PC
Geranium tuberosum L. subsp. ***tuberosum*** - uncultivated land - PC
Geranium versicolor L. - cool woods - PC

Grossulariaceae

Ribes alpinum L. - *Fagus sylvatica* woods - PC
Ribes multiflorum Kit. ex Roem. & Schult. subsp. ***multiflorum*** - *Fagus sylvatica* wood glades, scrub and cool environments - PC
Ribes petraeum Wulfen - D - Barrea (Tenore 1835), between Riotorto and Barrea (Tenore and Gussone 1842).
Ribes rubrum L. - D - Bocche di Chiarano (Gravina 1812; Tenore 1831).
Ribes uva-crispa L. subsp. ***uva-crispa*** - *Fagus sylvatica* wood glades, scrub and hedges, to 1,600 m

H

Hydrangeaceae

*A ***Philadelphus coronarius*** L. - uncultivated land - NAT - Picinisco, Valle del Melfa below the village!

Hypericaceae

Hypericum androsaemum L. - cool woods - R - Fiume Mollarino, sotto S. Biagio! (Conti 1995), Valle del Melfa below Picinisco! The report of *H. hircinum* for Valle del Melfa (Terracciano 1873) is probably to be referred to this taxon.

Hypericum hirsutum L. - humid meadows - PC - The report of *H. tomentosum* for Valle Canneto (Terracciano 1873) is to be referred to this taxon (Grande 1924).

Hypericum hyssopifolium Chaix - humid stony slopes - R - Rif. Pesco di Iorio, territory of Villavallelonga (Anzalone and Bazzichelli 1960).

Hypericum montanum L. - open woods - C

Hypericum perfoliatum L. - scrub, cool grassy land - NC - territory of Villavallelonga (Grande 1904).

Hypericum perforatum L. subsp. ***perforatum*** - fields, arid meadows - C

Hypericum perforatum L. subsp. ***veronense*** (Schrank) A. Fröhl. (*H. perforatum* L. var. *angustifolium* Borkh.; *H. perforatum* L. var. *microphyllum* DC.; *H. veronense* Schrank) - fields, arid meadows - C

Hypericum richeri Vill. subsp. ***richeri*** - stony slopes, pastures, scrub - PC

Hypericum tetrapterum Fr. (*H. quadrangulum* L.) - marshy environments, banks - C

F. Conti, F. Bartolucci, *The Vascular Flora of the National Park of Abruzzo, Lazio and Molise (Central Italy)*, Geobotany Studies, DOI 10.1007/978-3-319-09701-5_14

I

Iridaceae

Chamaeiris foetidissima (L.) Medik. (*Iris foetidissima* L.; *Xiphion foetidissimum* (L.) Parl.) - cool woods - RR - near S. Gennaro! (Conti 1995) - The nomenclature according to Peruzzi et al. (2014a).

****Crocus biflorus*** Mill. - meadows - RR - Rivoli!

Crocus neapolitanus (Ker Gawl.) Loisel. (*C. vernus* auct. Fl. Ital.; *C. vernus* (L.) Hill var. *neapolitanus* Ker Gawl.) - montane pastures - CC

Crocus reticulatus Steven ex Adams subsp. ***reticulatus*** - arid pastures - R - Monti della Marsica (Anzalone and Bazzichelli 1960 from specimens collected by Grande)

Gladiolus illyricus Koch - fields, uncultivated land - R - near the Abbazia of Castel S. Vincenzo!, slopes of Monte Mattone, over Pizzone! (Conti 1995).

Gladiolus italicus Mill. (*G. segetum* Ker Gawl.; *G. spathaceus* Parl.) - fields, uncultivated land - PC

A ***Iris germanica*** L. (incl. *I. florentina* L.) - uncultivated land - NAT

E ***Iris marsica*** I. Ricci & Colas. (Fig. 1) - arid pastures - PC

Limniris pseudacorus (L.) Fuss. (*Iris pseudacorus* L.; *Xiphion pseudacorus* (L.) Schrank) - riversides, marshy environments - R - loc. Il Pantano (Pedrotti 1983), Cartiera! (Conti 1992), Montenero Valcocchiara (Conti 1995) - The nomenclature according to Peruzzi et al. (2014a).

Romulea bulbocodium (L.) Sebast. & Mauri (*Crocus bulbocodium* L.) - arid meadows - RR - near the Abbazia of Castel S. Vincenzo! (Conti 1995).

F. Conti, F. Bartolucci, *The Vascular Flora of the National Park of Abruzzo, Lazio and Molise (Central Italy)*, Geobotany Studies, DOI 10.1007/978-3-319-09701-5_15

Fig. 1 *Iris marsica* (Photo by F. Conti)

J

Juglandaceae

A ***Juglans regia*** L. - watercourses - NAT

Juncaceae

Juncus articulatus L. - marshes, ditches, humid meadows - C

Juncus bufonius L. - humid environments - PC

Juncus bulbosus L. - NC - at Forestella below Monte Forcellone (Tenore and Gussone 1842).

Juncus compressus Jacq. - humid environments - PC

Juncus effusus L. subsp. ***effusus*** (*J. effusus* L. subsp. *fistulosus* (Guss.) Cif. & Giacom.; *J. fistulosus* Guss.) - marshes, humid meadows - PC

Juncus inflexus L. subsp. ***inflexus*** (*J. depauperatus* Ten.) - pools and watercourses - C

Juncus striatus Schousb. ex E. Mey. - D - Villetta Barrea, Pescasseroli (Anzalone and Bazzichelli 1960).

Luzula campestris (L.) DC. - pastures - C

Luzula congesta (Thuill.) Lej. (*Juncus congestus* Thuill.; *L. multiflora* (Retz.) Lej. subsp. *congesta* (Thuill.) Hyl.) - NC - Valico di Pantano (Vaccari and Wilczek 1940).

Luzula forsteri (Sm.) DC. (*Juncus forsteri* Sm.) - woods - C

Luzula multiflora (Ehrh.) Lej. - pastures - PC

Luzula pilosa (L.) Willd. (*Juncus pilosus* L.) - D - faggete chiuse (Bruno and Bazzichelli 1966), Settefrati (Petriglia 2004).

F. Conti, F. Bartolucci, *The Vascular Flora of the National Park of Abruzzo, Lazio and Molise (Central Italy)*, Geobotany Studies, DOI 10.1007/978-3-319-09701-5_16

Luzula spicata (L.) DC. subsp. ***bulgarica*** (Chrtek & Křísa) Gamisans (*L. bulgarica* Chrtek & Křísa) - meadows at high altitude - C

E ***Luzula sylvatica*** (Huds.) Gaudin subsp. ***sicula*** (Parl.) K. Richt. (*L. sieberi* Tausch subsp. *sicula* (Parl.) Parl., nom. inval.; *L. sicula* Parl.) - open woods, scrub, montane pastures - PC

Luzula sylvatica (Huds.) Gaudin subsp. ***sieberi*** (Tausch) K. Richt. (*L. sieberi* Tausch) - open woods, scrub, montane pastures - PC

Luzula sylvatica (Huds.) Gaudin subsp. ***sylvatica*** - woods - PC

****Luzula taurica*** (V. I. Krecz.) Novikov - pastures - R - Casone del Medico!, Le Forme! New species for Molise region.

Oreojuncus monanthos (Jacq.) Záv. Drábk. & Kirschner (*Juncus monanthos* Jacq.; *J. trifidus* L. subsp. *monanthos* (Jacq.) Asch. & Graebn.) - cool meadows at high altitude - C

Juncaginaceae

Triglochin palustris L. (Fig. 1) - peaty meadows - RR - La Camosciara! (Conti 1995).

Fig. 1 *Triglochin palustris* (Photo by F. Conti)

L

Lamiaceae

Ajuga chamaepitys (L.) Schreb. subsp. ***chamaepitys*** - uncultivated land, arid meadows - C

Ajuga chamaepitys (L.) Schreb. subsp. *chia* (Schreb.) Arcang. (*A. chia* Schreb.) - NC - Picinisco (Tenore 1835; Tenore and Gussone 1842).

Ajuga pyramidalis L. - arid meadows - RR - Montagnola (Hennecke and Hennecke 1999).

Ajuga reptans L. - woods, glades - C

E ***Ajuga tenorei*** C. Presl (Fig. 1) - stony pastures and *Nardus* grassland - PC

Ballota nigra L. subsp. ***meridionalis*** (Bég.) Bég. (*B. nigra* L. var. *foetida* Vis.; *B. velutina* Posp.; *B. nigra* L. subsp. *velutina* (Posp.) Patzak) - ruderal environments - C

E ***Betonica alopecuros*** L. subsp. ***divulsa*** (Ten.) Bartolucci & Peruzzi (*Betonica divulsa* Ten.; *Stachys alopecuros* (L.) Benth. subsp. *divulsa* (Ten.) Grande) - stony montane pastures - PC - The nomenclature according to Bartolucci et al. (2014b).

Betonica officinalis L. (*B. serotina* Host; *Stachys officinalis* (L.) Trevis.; *St. officinalis* (L.) Trevis. subsp. *serotina* (Host) Murb.) - pastures, open woods - C - the reports of *S. alpina* are to be referred to this taxon. The nomenclature according to Bartolucci et al. (2014b).

Betonica hirsuta L. (*B. pradica* Zanted.; *Stachys pradica* (Zanted.) Greuter & Pignatti) - D - Monte Meta (Terracciano 1872). Most probably to be excluded. The nomenclature according to Bartolucci et al. (2014b).

Clinopodium acinos (L.) Kuntze subsp. ***acinos*** (*Acinos arvensis* (Lam.) Dandy subsp. *arvensis*) - arid pastures - PC

F. Conti, F. Bartolucci, *The Vascular Flora of the National Park of Abruzzo, Lazio and Molise (Central Italy)*, Geobotany Studies, DOI 10.1007/978-3-319-09701-5_17

Fig. 1 *Ajuga tenorei* (Photo by F. Conti)

Clinopodium alpinum (L.) Merino subsp. ***meridionale*** (Nyman) Govaerts (*Calamintha alpina* (L.) Lam. subsp. *meridionalis* Nyman; *S. alpina* (L.) Scheele subsp. *meridionalis* (Nyman) Greuter & Burdet; *Acinos alpinus* (L.) Moench subsp. *meridionalis* (Nyman) P. W. Ball) - arid and stony pastures - C - a critical taxonomic revision of *C. alpinum* complex is necessary.

Clinopodium grandiflorum (L.) Stace (*Satureja grandiflora* (L.) Scheele; *Calamintha grandiflora* (L.) Moench) - woods - PC

Clinopodium graveolens (M. Bieb.) Kuntze (*Acinos graveolens* (M. Bieb.) Link; *Satureja graveolens* (M. Bieb.) Caruel; *Acinos rotundifolius* auct.; *Clinopodium rotundifolium* auct.; *Thymus graveolens* M. Bieb.) - NC - Villavallelonga (Grande 1904), Collelongo (Grande 1912).

Clinopodium nepeta (L.) Kuntze subsp. ***nepeta*** (*Calamintha nepeta* (L.) Savi subsp. *nepeta*) - arid meadows, ruderal environments - PC

Clinopodium nepeta (L.) Kuntze subsp. ***spruneri*** (Boiss.) Bartolucci & F. Conti (*Calamintha glandulosa* (Req.) Benth.; *C. nepeta* (L.) Savi subsp. *glandulosa* (Req.) P. W. Ball; *C. nepeta* (L.) Savi subsp. *spruneri* (Boiss.) Nyman; *C. spruneri* Boiss.; *Thymus glandulosus* Req.) - arid meadows, ruderal environments - R - Gole del Sagittario! (Conti and Tinti 2012).

Clinopodium nepeta (L.) Kuntze subsp. ***sylvaticum*** (Bromf.) Peruzzi & F. Conti (*Calamintha menthifolia* Host; *C. sylvatica* Bromf. subsp. *sylvatica*; *C. nepeta* (L.) Savi subsp. *sylvatica* (Bromf.) R. Morales; *Satureja menthifolia* (Host) Fritsch) - open woods, hedges - PC

Clinopodium suaveolens (Sm.) Kuntze (*Satureja suaveolens* (Sm.) Watzl - Zeman; *Acinos suaveolens* (Sm.) Loudon; *Thymus suaveolens* Sm.) - NC - surroundings of Gioia Vecchio, from Barrea to Villetta Barrea (Vaccari and Wilczek 1940).

Clinopodium vulgare L. subsp. ***vulgare*** (*Satureja vulgaris* (L.) Fritsch subsp. *vulgaris*) - open woods - PC

Galeopsis angustifolia Ehrh. ex Hoffm. subsp. ***angustifolia*** - stony areas, arid uncultivated land - C

Galeopsis ladanum L. - stony areas, stony pastures - PC

Galeopsis pubescens Besser (*G. murriana* Borbás & Wettst. ex Murr) - hedges, open woods - PC

Galeopsis tetrahit L. - fields, screes - PC - the report of *G. bifida* for Picinisco (Petriglia 2004) is to be referred to this taxon.

Glechoma hederacea L. - NC - Picinisco (Tenore and Gussone 1842), generically for the Parco (Rovesti and Rovesti 1934), Villavallelonga (Anzalone and Bazzichelli 1960 quote Grande).

Glechoma hirsuta Waldst. & Kit. (*G. hederacea* L. subsp. *hirsuta* (Waldst. & Kit.) Gams) - hedges, open woods - PC

Hyssopus officinalis L. subsp. ***aristatus*** (Godr.) Nyman (*H. officinalis* L. subsp. *pilifer* (Pant.) Murb.) - arid and stony pastures - PC

Lamium album L. subsp. ***album*** - uncultivated land, nitrified environments - PC

Lamium amplexicaule L. - fields, ruderal environments - C

Lamium bifidum Cirillo subsp. ***bifidum*** - open woods, uncultivated land, pastures - R - Passo di Croci between Scanno and Roccapia (Mennema 1989), Vallone Lacerno (Conti 1995).

Lamium flexuosum Ten. - stony areas, open woods - PC

Lamium galeobdolon (L.) L. subsp. ***galeobdolon*** (*Lamiastrum galeobdolon* (L.) Ehrend. & Polatschek subsp. *galeobdolon*) - *Fagus sylvatica* woods - PC

Lamium garganicum L. subsp. ***laevigatum*** Arcang. - cliffs, open woods, screes - C

Lamium garganicum L. subsp. ***striatum*** (Sm.) Hayek (*L. garganicum* L. subsp. *gracile* (Briq.) Greuter & Burdet) - cliffs, screes - PC

Lamium maculatum L. - scrub, hedges, uncultivated land - CC

Lamium purpureum L. (*L. hybridum* Vill.) - fields, ruderal environments - C

Lycopus europaeus L. (*L. europaeus* L. subsp. *mollis* (A. Kern.) Rothm. ex Skalický; *L. europaeus* L. subsp. *menthaefolius* (Mabille) Skalický) - ditches, riversides, humid meadows - R - Picinisco (Tenore and Gussone 1842), riversides of F. Volturno near Cartiera! (Conti 1995).

Marrubium incanum Desr. - pastures, uncultivated areas - PC

Marrubium vulgare L. - ruderal environments, uncultivated land - PC

Melissa officinalisL. (incl. *M. altissima* Sm.; incl. *M. romana* Mill.; incl. *M. officinalis* L. subsp. *altissima* (Sm.) Arcang.) - ruderal environments, hedges - C

Melittis melissophyllum L. subsp. ***melissophyllum*** - woods and margins - C

Mentha aquatica L. subsp. ***aquatica*** - riversides, humid uncultivated land - PC

Mentha longifolia (L.) Huds. - riversides, humid meadows - C

Mentha microphylla K. Koch (*Mentha spicata* L. subsp. *condensata* (Briq.) Greuter & Burdet) - riversides, humid meadows - R - Villavallelonga (Grande

1924), Fonte Cupa near Castel S. Vincenzo!, Gole del Sagittario! (Conti and Tinti 2012).

Mentha pulegium L. subsp. ***pulegium*** - clayey and humid fields, riversides - PC

Mentha spicata L. (incl. *M. spicata* L. subsp. *glabrata* (Lej. & Courtois) Lebeau; *M. viridis* (L.) L.; *M. spicata* L. var. *viridis* L.) - riversides, humid uncultivated land - PC

Mentha suaveolens Ehrh. subsp. ***suaveolens*** - ditches, humid environments - R - F. Mollarino (Zodda 1931), Capo Volturno (Buchwald 1995).

E ***Micromeria graeca*** (L.) Benth. ex Rchb. subsp. ***tenuifolia*** (Ten.) Nyman (*Satureja graeca* L. subsp. *tenuifolia* (Ten.) Arcang.; *S. tenuifolia* Ten.; *S. graeca* L. var. *tenuifolia* (Ten.) Vis.) - arid and stony pastures, cliffs - C - the report of *M. marginata* for Rif. Aremogna (Hennecke and Hennecke 1999 sub *Satureja piperella*) is probably to be referred to this taxon.

Micromeria juliana (L.) Benth. ex Rchb. (*Satureja juliana* L.) - arid and stony pastures, cliffs - PC

Nepeta cataria L. - ruderal environments - PC

Nepeta nuda L. subsp. ***nuda*** - margins of woods, pastures - PC

Origanum vulgare L. subsp. ***vulgare*** - open woods, scrub - PC

Phlomis fruticosa L. (Fig. 2) - arid pastures, garigue - RR - generically recorded for the Park (Rovesti and Rovesti 1934) and near Pescina outside the Park but close to the edge of the external buffer zone (Conti 1995).

Fig. 2 *Phlomis fruticosa* (Photo by F. Conti)

Phlomis herba-venti L. subsp. *herba-venti* - NC - generically recorded for the Park (Rovesti and Rovesti 1934) and observed near Pescina outside the Park but close to the buffer external zone.

Prunella laciniata (L.) L. - arid pastures - PC

Prunella vulgaris L. subsp. ***vulgaris*** - open woods, cool meadows - C

A ***Rosmarinus officinalis*** L. - ruderal environments - CAS

Salvia aethiopis L. - arid pastures - RR - near Rivoli! It has been recently recorded for Massa d'Albe (Conti et al. 2011c) and thus confirmed to the Flora of Abruzzo.

Salvia argentea L. - arid pastures - PC

Salvia clandestina L. - arid uncultivated areas, sandy alluvial soils - RR - between Pescina and Ortona dei Marsi! (Conti and Minutillo 2001).

Salvia glutinosa L. - cool woods - PC

Salvia nemorosa L. subsp. ***nemorosa*** (*S. sylvestris* auct., non L.) - montane pastures - RR - Frattura! (Conti 1998).

Salvia officinalis L. (incl. *S. officinalis* L. var. *angustifolia* Ten.) - stony slopes, arid meadows - R - Vallone di S. Onofrio (Viegi et al. 1990), Serra Lunga! (Pirone and Tammaro 1997).

Salvia pratensis L. subsp. ***pratensis*** - uncultivated land, arid meadows - RR - between Gioia dei Marsi and Gioia Vecchio (Anzalone and Bazzichelli 1960).

Salvia sclarea L. - arid pastures, uncultivated areas - PC

Salvia verbenaca L. (*S. multifida* Sm.) - arid meadows - C

Salvia virgata Jacq. - pastures, arid uncultivated areas - PC

Satureja montana L. subsp. ***montana*** - thermophilous stony slopes - CC

Scutellaria alpina L. subsp. ***alpina*** - stony slopes at high altitudes - PC

Scutellaria altissima L. - woods - PC

Scutellaria columnae All. subsp. ***columnae*** - scrub, mixed woods - C - the report of *S. columnae* subsp. *gussonei* (Petriccione et al. 1994) is to be referred to this taxon.

Scutellaria galericulata L. - humid meadows - RR - La Brionna, il Pantano (Tammaro 1986).

Stachys annua (L.) L. subsp. *annua* (*St. pubescens* Ten.) - NC - Valico di Pantano (Vaccari and Wilczek 1940).

Stachys germanica L. subsp. ***salviifolia*** (Ten.) Gams (*St. cretica* L. subsp. *salviifolia* (Ten.) Rech. f.; *St. salviifolia* Ten.) - arid pastures - C

Stachys heraclea All. - pastures - PC

E ***Stachys italica*** Mill. (*Sideritis italica* (Mill.) Greuter & Burdet, *S. brutia* Ten.; *S. sicula* Ucria; *S. syriaca* auct. Fl. Ital.) - arid and stony pastures - C - The nomenclature according to Bartolucci et al. (2014b).

Stachys montana (L.) Peruzzi & Bartolucci subsp. ***montana*** (*Sideritis montana* L. subsp. *montana* - arid meadows - PC - The nomenclature according to Bartolucci et al. (2014b).

Stachys recta L. subsp. ***grandiflora*** (Caruel) Arcang. (*St. recta* L. var. *grandiflora* Caruel; *St. recta* L. subsp. *labiosa* (Bertol.) Briq.) - montane stony slopes - PC - the reports of *S. recta* subsp. *subcrenata* for Mainarde (Conti 1995) are to be referred to this taxon.

Stachys recta L. subsp. ***recta*** - thermophilous, arid and stony pastures - PC

Stachys romana (L.) E. H. L. Krause subsp. ***romana*** (*Sideritis romana* L. subsp. *romana*) - arid meadows - C - The nomenclature according to Bartolucci et al. (2014b).

Stachys sylvatica L. - woods, cool glades - PC

Stachys tymphaea Hausskn. - pastures - C

Teucrium botrys L. - uncultivated land - RR - near Rocca Altiera, loc. Guado Sambuco! (Conti 1995).

Teucrium capitatum L. subsp. ***capitatum*** (*T. polium* L. subsp. *capitatum* (L.) Arcang.) - arid and stony meadows - PC

Teucrium chamaedrys L. subsp. ***chamaedrys*** - arid pastures, glades - C

Teucrium flavum L. subsp. ***flavum*** - thermophilous stony slopes - R - Picinisco allo Schioppaturo (Tenore and Gussone 1842), Gole del Sagittario! (Anzalone 1961; Conti and Tinti 2012), Monte la Rocca near Picinisco (Spada 1979), Monte Mattone near Pizzone! (Conti 1995).

Teucrium montanum L. - stony slopes, stony pastures - C

Teucrium scordium L. subsp. ***scordioides*** (Schreb.) Arcang. (*T. scordioides* Schreb.) - humid meadows, marshes - R - Picinisco alla Cartiera and allo Schioppaturo (Tenore and Gussone 1842), "Valvori presso una sorgente nell'andare a S. Biagio" (Terracciano 1873), loc. Il Pantano (Pedrotti 1983).

Teucrium siculum (Raf.) Guss. subsp. ***siculum*** (*T. scorodonia* L. subsp. *crenatifolium* (Guss.) Arcang.; *T. scorodonia* L. var. *crenatifolium* Guss.) - woods - RR - near Castel S. Vincenzo! (Paura and Abbate 1995; Conti and Minutillo 1998). The record of *T. scorodonia* for Valle di Canneto is probably to be referred to this taxon.

Thymus longicaulis C. Presl subsp. ***longicaulis*** - margins of wood, glades - C - The records of *Th. pulegioides* are mainly to be referred to this taxon (Bartolucci 2010).

Thymus oenipontanus Heinr. Braun (*Th. decipiens* Heinr. Braun; *Th. pannonicus* auct. Fl. Ital. p.p.) - arid and stony pastures - C

Thymus praecox Opiz subsp. ***polytrichus*** (A. Kern. ex Borbás) Jalas - pastures at high altitude - C

Thymus striatus Vahl subsp. ***acicularis*** (Waldst. & Kit.) Ronniger (*Th. acicularis* Waldst. & Kit.) - arid and stony pastures - PC - this taxon has been recently revaluated (Bartolucci and Peruzzi 2014).

A ***Thymus vulgaris*** L. subsp. ***vulgaris*** - arid and stony pastures - CAS

Lauraceae

Laurus nobilis L. - cool deep valleys between the Mediterranean and supramediterranean horizon - R - Valle del Melfa! (Spada 1979), foce del F. Mollarino (Zodda 1931), between Anversa and Colle del Tuppo! (Conti and Tinti 2012).

Lentibulariaceae

E ***Pinguicula vallis-regiae*** F. Conti & Peruzzi (Fig. 3) - wet cliffs, humid and peaty meadows - RR - Camosciara! (Conti and Peruzzi 2006).

Liliaceae

Fritillaria montana Hoppe ex W. D. J. Koch (*F. orientalis* auct. Fl. Ital.; *F. tenella* auct. Fl. Ital.; *F. orsiniana* Parl.) - pastures - PC

Gagea bohemica (Zauschn.) Schult. & Schult. f. (*G. bohemica* (Zauschn.) Schult. & Schult. f. subsp. *saxatilis* (Mert. & W. D. J. Koch) Asch. and Graebn.; incl.

Fig. 3 *Pinguicula vallis-regiae* (Photo by F. Conti)

L

Fig. 4 *Gagea ramulosa* (Photo by F. Bartolucci)

G. busambarensis (Tineo) Parl.; *Ornithogalum bohemicum* Zauschn.; *O. busambarense* Tineo) - arid and stony meadows - C

Gagea fragifera (Vill.) Ehr. Bayer & G. López (*G. fistulosa* (Ramond ex DC.) Ker. Gawl.; *Ornithogalum fragiferum* Vill.) - cool meadows - PC

Gagea lacaitae A. Terracc. - arid pastures - PC - The previous records of *G. granatelli* from the Park are to be referred to this species (Tison et al. 2012).

Gagea lutea (L.) Ker Gawl. (*Ornithogalum luteum* L.) - *Fagus sylvatica* woods - PC

Gagea minima (L.) Ker Gawl. (*Ornithogalum minimum* L.) - arid pastures - RR - Prati di Mezzo (Picinisco) (Minutillo 1995).

Gagea pratensis (Pers.) Dumort. (*G. pomeranica* Ruthe; *Ornithogalum pratense* Pers.) - arid meadows - RR - Valle del Giovenco near Rivoli! (Bartolucci and Peruzzi 2007; Conti et al. 2008).

Gagea ramulosa A. Terracc. (*G. dubia* auct. Fl. Ital.) (Fig. 4) - arid meadows - RR - Valle del Giovenco near Pescina! (Peruzzi and Bartolucci 2006; Bartolucci and Peruzzi 2007).

Gagea villosa (M. Bieb.) Sweet (*G. arvensis* auct.; *Ornithogalum villosum* M. Bieb.) - fields, arid uncultivated land - PC

Lilium bulbiferum L. subsp. ***croceum*** (Chaix) Jan (*L. croceum* Chaix; *L. bulbiferum* L. var. *croceum* (Chaix) Pers.) - glades, margins of woods - C

Lilium martagon L. - glades, open *Fagus sylvatica* woods, subalpine pastures - C

Streptopus amplexifolius (L.) DC. (*Uvularia amplexifolia* L.) - cool *Fagus sylvatica* woods, in the impluviums - PC

****Tulipa pumila*** Moench (*T. australis* Link, *T. sylvestris* L. subsp. *australis* (Link) Pamp.) (Fig. 5) - montane pastures - RR - loc. Canale near Collelongo! The nomenclature according to Christenhusz et al. (2013).

Fig. 5 *Tulipa pumila* (Photo by F. Conti)

Linaceae

Linum alpinum Jacq. (*L. alpinum* Jacq. subsp. *gracilius* auct.; *L. alpinum* Jacq. subsp. *julicum* (Hayek) Hegi; *L. perenne* L. subsp. *alpinum* (Jacq.) Stoj. and Stef.; *L. julicum* Hayek) - montane and subalpine meadows - PC

Linum bienne Mill. - uncultivated land, fields - PC

Linum capitatum Kit. ex Schult. subsp. ***serrulatum*** (Bertol.) Hartvig (Fig. 6) - stony pastures at high altitude - PC

Linum catharticum L. subsp. ***catharticum*** - stony humid meadows - PC

Linum catharticum L. subsp. ***suecicum*** (Murb. ex Hayek) Hayek - stony humid meadows - RR - Camosciara! (Ballelli et al. 2005).

Linum corymbulosum Rchb. (*L. strictum* L. subsp. *corymbulosum* (Rchb.) Rouy) - arid meadows - PC

Linum narbonense L. - NC - Villavallelonga (Grande 1916).

Linum nodiflorum L. - NC - Picinisco (Tenore and Gussone 1842).

Linum strictum L. subsp. ***strictum*** - arid meadows - R - between Gioia dei Marsi and Gioia Vecchio (Vaccari and Wilczek 1940), near the Abbazia of Castel S. Vincenzo! (Conti 1995).

Linum tenuifolium L. - arid meadows - C

Linum tommasinii (Rchb.) Nyman (*L. collinum* Guss.; *L. tommasinii* Rchb.; *L. austriacum* L. subsp. *tommasinii* (Rchb.) Greuter & Burdet; *L. alpinum* Jacq. subsp. *gracilius* Pignatti) - arid meadows - PC

*A ***Linum usitatissimum*** L. - uncultivated land, fields - NAT - S. Michele a Foce!

Linum viscosum L. - hill and montane pastures - R - near lago di Castel S. Vincenzo! (Conti 1995), Balze di Marcello!

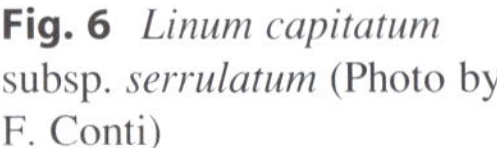

Fig. 6 *Linum capitatum* subsp. *serrulatum* (Photo by F. Conti)

Loranthaceae

Loranthus europaeus Jacq. - hemiparasite on *Quercus pubescens* and occasionally on *Q. cerris* and *Castanea sativa*, to 1,000 m - PC

Lythraceae

Lythrum hyssopifolia L. - marshy environments, muds - R - Picinisco at Cartiera (Tenore and Gussone 1842), F. Volturno near Castel S. Vincenzo! (Conti 1995).

Lythrum salicaria L. - borders of the ditches, watercourses, marshes - C

Peplis portula L. (*Lythrum portula* (L.) D. A. Webb) - marshy environments - R - Monte Forcellone (Conti 1995), Prato di Mezzo!

M

Malvaceae

A ***Alcea biennis*** Winterl subsp. ***biennis*** (*A. pallida* (Willd.) Waldst. & Kit.; *Althaea pallida* Willd.) - arid uncultivated land - NAT

A ***Alcea rosea*** L. (*Althaea rosea* (L.) Cav.) - arid uncultivated land - CAS

Althaea cannabina L. - humid uncultivated land - R - Picinisco (Tenore and Gussone 1842), Monte Castelnuovo near the village! (Conti 1995).

Althaea officinalis L. (*A. taurinensis* DC.) - riversides and ditches - RR - sources of Volturno river! (Conti 1995).

Malva alcea L. - uncultivated land - PC

Malva moschata L. - uncultivated land - C

Malva neglecta Wallr. - ruderal environments, uncultivated land - C

Malva nicaeensis All. - NC - source of Melfa (Terracciano 1873).

Malva parviflora L. - NC - surroundings of Gioia Vecchio (Vaccari and Wilczek 1940).

Malva setigera Schimp. & Spenn. (*Althaea hirsuta* L.) - arid pastures, uncultivated areas - C

Malva sylvestris L. (*M. ambigua* Guss.; *M. sylvestris* L. subsp. *ambigua* (Guss.) P. Fourn.) - ruderal environments, pastures - CC

Malva thuringiaca (L.) Vis. (*Lavatera ambigua* DC.; *L. thuringiaca* L. subsp. *ambigua* (DC.) Nyman; *L. thuringiaca* L. subsp. *thuringiaca*) - margins of woods - PC

Tilia cordata Mill. - mesophilous woods - PC

Tilia platyphyllos Scop. subsp. ***cordifolia*** (Besser) C. K. Schneid. (*T. cordifolia* Besser) - mesophilous woods - PC

****Tilia platyphyllos*** Scop. subsp. ***pseudorubra*** C. K. Schneid. - mesophilous woods - RR - Scanno, loc. Imposto! New taxon for Abruzzo region.

F. Conti, F. Bartolucci, *The Vascular Flora of the National Park of Abruzzo, Lazio and Molise (Central Italy)*, Geobotany Studies, DOI 10.1007/978-3-319-09701-5_18

Melanthiaceae

Paris quadrifolia L. - cool *Fagus sylvatica* woods - PC

Veratrum album L. (*V. album* L. subsp. *lobelianum* (Bernh.) Arcang.; *V. lobelianum* Bernh.) - glades, pastures and meadows at high altitude - PC

Veratrum nigrum L. - open woods, montane pastures - PC

Menyanthaceae

Menyanthes trifoliata L. (Fig. 1) - montane marshy environments - R - Il Lagozzo! (Naviglio 1984), Passo Godi!, Campitelli! (Conti 1998), Montenero Val Cocchiara!

Moraceae

Ficus carica L. - cliffs, walls - PC

Fig. 1 *Menyanthes trifoliata* (Photo by F. Conti)

O

Oleaceae

Fraxinus angustifolia Vahl subsp. ***oxycarpa*** (Willd.) Franco & Rocha Afonso (*F. oxycarpa* Willd.) - ditches, watercourses, humid woods - R - Cartiera!, Sorgenti del Volturno! (Conti 1992), between Pizzone and Castel S. Vincenzo!, Pescasseroli!

Fraxinus excelsior L. subsp. ***excelsior*** - cool woods - PC

Fraxinus ornus L. subsp. ***ornus*** - scrub and mixed woods - CC

A ***Jasminum officinale*** L. - stony slopes - CAS

A *Ligustrum lucidum* W. T. Aiton - CAS

Ligustrum vulgare L. - hedges, scrub, open broadleaf woods and margins - PC

A ***Olea europaea*** L. subsp. ***europaea*** (*O. europaea* L. subsp. *oleaster* (Hoffmanns. and Link) Negodi) - stony slopes - NAT

Phillyrea latifolia L. - maquis, *Quercus ilex* woods, broadleaf woods of the supramediterranean to 800 m - PC

*A ***Syringa vulgaris*** L. - ruderal environments - CAS - Camosciara near the piazzale!, Piana di Pescasseroli!

Onagraceae

Chamaenerion angustifolium (L.) Scop. (*Epilobium angustifolium* L.) - glades, screes, humid depressions - PC

Chamaenerion dodonaei (Vill.) Schur ex Fuss (*Ch. rosmarinifolium* (Haenke) Moench; *Epilobium rosmarinifolium* Haenke; *E. dodonaei* Vill.) - unstable stony slopes - R - Monte Marrone! (Conti 1995), Settefrati (Val Canneto near the source of Canneto) (Petriglia 2004).

F. Conti, F. Bartolucci, *The Vascular Flora of the National Park of Abruzzo, Lazio and Molise (Central Italy)*, Geobotany Studies, DOI 10.1007/978-3-319-09701-5_19

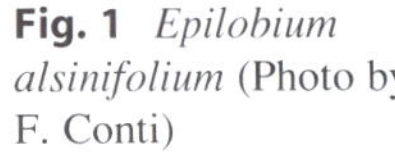

Fig. 1 *Epilobium alsinifolium* (Photo by F. Conti)

Circaea lutetiana L. subsp. ***lutetiana*** - cool woods - PC
Epilobium alpestre (Jacq.) Krock. - humid meadows - RR - Le Forme! (Conti 1995).
Epilobium alsinifolium Vill. (Fig. 1) - sources, shady and wet cliffs - R - Camosciara (Fiori 1927), Valle Venafrana between Monte Cavallo and Monte a Mare! (Conti 1995).
Epilobium hirsutum L. - humid environments - CC
Epilobium montanum L. - *Fagus sylvatica* woods - CC
Epilobium palustre L. - marshy environments - R - Villetta Barrea (Anzalone and Bazzichelli 1960), Pantano di Scanno (Avena and Rosati 1974), Campitelli! (Conti 1995).
Epilobium parviflorum Schreb. - humid environments - C
Epilobium tetragonum L. subsp. ***tetragonum*** - humid environments - PC

Ophioglossaceae

Botrychium lunaria (L.) Sw. (*Osmunda lunaria* L.) - high meadows - C
Ophioglossum vulgatum L. subsp. ***vulgatum*** - humid meadows - R - Il Lagozzo! (Conti 1994), Padura between Pescasseroli and Bisegna (Conti 1998), Valle del Templo!

Orchidaceae

Anacamptis coriophora (L.) R. M. Bateman, Pridgeon & M. W. Chase (*Orchis fragrans* Pollini; *O. coriophora* L.) - arid meadows - R - Vallone dal Carapale at Scanno (Gravina 1812), Valleporcina, Balzo di Canneto - Canneto (Conti 1995).

Anacamptis laxiflora (Lam.) R. M. Bateman, Pridgeon & M. W. Chase (*Orchis laxiflora* Lam.) (Fig. 2) - humid meadows - R - sources of Volturno (Conti 1992), Montenero Val Cocchiara (Hennecke and Hennecke 1999).

Anacamptis morio (L.) R. M. Bateman, Pridgeon & M. W. Chase (*Orchis picta* Loisel.; *O. morio* L.) - arid meadows - CC

Anacamptis pyramidalis (L.) Rich. - arid meadows - CC

Barlia robertiana (Loisel.) Greuter (*B. longibracteata* (Biv.) Parl.; *Himantoglossum longibracteatum* (Biv.) Schltr.; *H. robertianum* (Loisel.) P. Delforge; *Orchis longibracteata* Biv.; *O. robertiana* Loisel.) - NC - Scanno (Tenore 1831).

Cephalanthera damasonium (Mill.) Druce (*C. alba* (Crantz) Simonk.; *C. pallens* Rich.; *Serapias damasonium* Mill.) - open woods, glades to about 1,700 m - C

Cephalanthera longifolia (L.) Fritsch (*C. ensifolia* Rich.) - woods to about 1,700 m - C

Fig. 2 *Anacamptis laxiflora* (Photo by F. Conti)

Fig. 3 *Cypripedium calceolus* (Photo by F. Conti)

Cephalanthera rubra (L.) Rich. (*C. maravignae* Tineo; *Serapias rubra* L.) - open woods, glades, scrub to about 1,800 m - C

Coeloglossum viride (L.) Hartm. (*Habenaria viridis* R. Br.; *Orchis viridis* Crantz; *Platanthera viridis* Lindl.; *Satyrium viride* L.; *Dactylorhiza viridis* (L.) R. M. Bateman, Pridgeon & Chase) - montane meadows from 800 to the tops - C

Corallorhiza trifida Châtel. (*C. innata* R. Br.) - *Fagus sylvatica* woods - PC

Cypripedium calceolus L. (Fig. 3) - clear areas in the *Fagus sylvatica* woods, between the *Fagus sylvatica* woods and the scrub with *Pinus mugo* - R - Camosciara! (Sipari 1926 quotes Pirotta; Fiori 1927; Pirotta 1933; Anzalone and Bazzichelli 1960), Val Fondillo! (Rovelli and Conti 1995).

Dactylorhiza incarnata (L.) Soó subsp. ***incarnata*** (Fig. 4) - humid meadows - PC

Dactylorhiza maculata (L.) Soó subsp. ***fuchsii*** (Druce) Hyl. (*D. fuchsii* (Druce) Soó; *Orchis fuchsii* (Druce) Hyl.; *O. maculata* L. subsp. *fuchsii* (Druce) Hyl.) - cool woods, humid meadows from 100 to about 2,000 m - C

Dactylorhiza maculata (L.) Soó subsp. ***saccifera*** (Brongn.) Diklić (*D. saccifera* (Brongn.) Soó; *O. maculata* L. subsp. *saccifera* (Brongn.) Soó; *O. saccifera* Brongn.) - cool woods, humid meadows - PC

Dactylorhiza sambucina (L.) Soó (*D. latifolia* (L.) H. Baumann & Künkele; *Orchis sambucina* L.) - montane pastures from 1,000 to about 2,200 m - C

Epipactis atrorubens (Hoffm. ex Bernh.) Besser (*E. atropurpurea* Raf.; *E. rubiginosa* (Crantz) W. D. J. Koch; *Helleborine atropurpurea* (Raf.) Schinz & Thell.) - open woods, glades, stony pastures from 1,000 to about 1,900 m - C

Epipactis helleborine (L.) Crantz subsp. ***helleborine*** - woods and glades from the hills to about 2,000 m - C

Epipactis helleborine (L.) Crantz subsp. ***latina*** W. Rossi & E. Klein - woods and glades - R - Barrea, Gioia dei Marsi, Opi - Forca d'Acero, Lago di Barrea, Villetta Barrea (Rossi and Klein 1987; Baumann and Baumann 1988; Baumann and Lorenz, 1988; Hoffmann 1989).

Epipactis helleborine (L.) Crantz subsp. ***orbicularis*** (K. Richt.) E. Klein (*E. distans* Arv.-Touv.) - woods and glades - RR - Gole del Sagittario! (Conti and Tinti 2012).

Fig. 4 *Dactylorhiza incarnata* subsp. *incarnata* (Photo by F. Conti)

Epipactis leptochila (Godfery) Godfery (incl. *E. leptochila* (Godfery) Godfery subsp. *neglecta* Kumpel) - *Fagus sylvatica* woods - R - Val Fondillo, Val Canneto (Conti 1995), Camosciara!

E ***Epipactis meridionalis*** H. Baumann & R. Lorenz (*E. gracilis* auct.) - *Fagus sylvatica* woods - RR - Monte Marrone (Hoffmann 1989).

Epipactis microphylla (Ehrh.) Sw. (*Helleborine microphylla* (Ehrh.) Schinz & Thell.; *Serapias microphylla* Ehrh.) - open woods, glades to about 1,500 m - C

Epipactis muelleri Godfery - woods - R - Valle di Mezzo (Steffan and Steffan 1986), Lago di Barrea (Baumann and Baumann 1988).

Epipactis palustris (L.) Crantz (*E. longifolia* All.; *Helleborine palustris* (L.) Schrank; *Serapias palustris* Mill.) - marshy environments - PC

Epipactis persica (Soó) Nannf. subsp. ***gracilis*** (B. Baumann & H. Baumann) W. Rossi (*E. gracilis* B. Baumann & H. Baumann, nom. illeg.; *E. persica* auct.; *E. baumanniorum* Soldano & F. Conti, nom. illeg.; *E. exilis* P. Delforge; *E. persica* (Soó) Nannf. subsp. *exilis* (P. Delforge) Kreutz) - *Fagus sylvatica* woods - R - Monti della Meta (Steffan and Steffan 1986), Monte La Rocca (Mainarde)! (Conti 1995), Valle Franchitta!

Epipactis purpurata Sm. (*E. viridiflora* Hoffm. ex Krock.; *E. violacea* (Dur.-Duq.) Boreau) - *Fagus sylvatica* woods - R - Valle Iannanghera (Conti 1995), La Meta (Betti 1997).

Fig. 5 *Epipogium aphyllum* (Photo by F. Conti)

Epipogium aphyllum Sw. (Fig. 5) - *Fagus sylvatica* woods - PC

Gymnadenia conopsea (L.) R. Br. (incl. *G. conopsea* (L.) R. Br. subsp. *densiflora* (Wahlenb.) K. Richt.; *Orchis conopsea* L.) - meadows to about 2,000 m - C

Himantoglossum adriaticum H. Baumann (Fig. 6) - arid meadows from the hills to about 1,500 m - PC - the reports of *H. hircinum* are to be referred to this taxon.

Limodorum abortivum (L.) Sw. (*Orchis abortiva* L.) - arid meadows from the hills to about 1,500 m - PC

Listera ovata (L.) R. Br. (*Ophrys ovata* L.) - cool woods from 600 to about 1,500 m - C

Neotinea tridentata (Scop.) R. M. Bateman, Pridgeon & M. W. Chase (*Orchis commutata* Tod.; *O. tridentata* Scop.) - arid meadows, spread from 500 to about 1,200 m - C

Neotinea ustulata (L.) R. M. Bateman, Pridgeon & M. W. Chase (*Orchis ustulata* L.) - meadows, spread from 800 to about 1,500 m - C

Neottia nidus-avis (L.) Rich. (*Ophrys nidus-avis* L.) - *Fagus sylvatica* woods - C

Nigritella widderi Teppner & E. Klein (*N. rubra* (Wettst.) K. Richt. subsp. *widderi* (Teppner & E. Klein) H. Baumann & R. Lorenz; *Gymnadenia widderi* (Teppner & E. Klein) Teppner & E. Klein) - meadows al high altitudes, over 1,700 m - PC

Ophrys apifera Huds. (*O. arachnites* Mill.; *O. rostrata* Ten.; incl. *O. apifera* Huds. var. *aurita* (Moggr.) Gremli; incl. *Ophrys apifera* Huds. var. *chlorantha*

Fig. 6 *Himantoglossum adriaticum* (Photo by F. Conti)

(Hegetschw.) Nyman; *O. sphegodes* Mill.; *O. holosericea* (F. W. Schimdt) Moench) - arid meadows to 1,200 m - C

E ***Ophrys appennina*** Romolini & Soca (*O. holosericea* auct. Fl. Ital. p.p.) - arid meadows - C

E ***Ophrys ausonia*** Devillers, Devillers-Tersch. & P. Delforge (*O. tommasinii* auct. Fl. Ital.) - arid meadows - R - Villetta Barrea, presso Scontrone (Griebl 2010).

Ophrys bertolonii Moretti subsp. ***bertolonii*** (*O. romolinii* Soca) - arid meadows - PC - the reports of *O. bertoloni* subsp. *bertoloniiformis* are to be referred to this taxon.

Ophrys classica Devillers-Tersch. and Devillers (*O. sphegodes* auct. Fl. Ital.) - arid meadows - CC

E ***Ophrys crabronifera*** Mauri (Fig. 7) (incl. *O. exaltata* Ten. subsp. *sundermannii* Soó; *O. fuciflora* (F. W. Schmidt) Moench subsp. *sundermannii* Soó; *O. argolica* H. Fleischm. ex Vierh. subsp. *crabronifera* (Sebast. & Mauri) Faurh.) - arid meadows - R - S. Michele a Foce!, Monte della Rocchetta!, Monte Mattone near Pizzone! (Conti 1992).

Ophrys dinarica Kranjčev & P. Delforge - arid meadows - PC - the reports of *O. apulica*, *O. scolopax* e *O. oestrifera* subsp. *bremifera* are probably to be referred to this taxon.

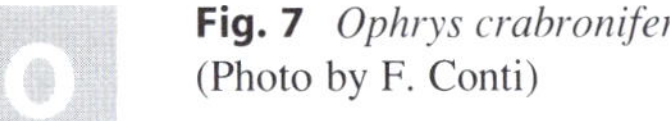

Fig. 7 *Ophrys crabronifera* (Photo by F. Conti)

Ophrys exaltata Ten. subsp. ***archipelagi*** (Gölz & H.R. Reinhard) Del Prete (Fig. 8) - arid meadows - RR - near Scontrone (Griebl 2010).

Ophrys funerea Viv. - arid meadows - PC

Ophrys gracilis (Büel, O. Danesch & E. Danesch) Paulus (*O. holosericea* (Burnm. f.) Greuter subsp. *gracilis* (Büel, O. Danesch & E. Danesch) Büel, O. Danesch & E. Danesch; *O. fuciflora* (F. W. Schmidt) Moench subsp. *gracilis* Büel, O. Danesch & E. Danesch) - arid meadows - RR - Passo del Diavolo (Hennecke and Hennecke 1999).

Ophrys incubacea Bianca (*O. atrata* Lindl., non L.) - arid meadows - PC

Ophrys insectifera L. (*O. muscifera* Huds.; *O. myodes* (L.) Jacq.; *O. muscaria* Lam.) - arid meadows - PC

Ophrys lacaitae Lojac. - arid meadows - R - near Alfedena (Conti 1995), Scapoli near Ponte Nuovo (Hennecke and Hennecke 1999).

E ***Ophrys lucana*** P. Delforge, Devillers-Tersch. & Devillers (*O. fusca* Link subsp. *lucana* (P. Delforge, Devillers-Tersch. & Devillers) Kreutz) - arid meadows - PC - the reports of *O. fusca* are to be referred to this taxon.

Ophrys passionis Sennen ex Devillers-Tersch. & Devillers subsp. ***passionis*** (*O. sphegodes* Mill. subsp. *garganica* E. Nelson, nom. inval.; *O. garganica* O. Danesch & E. Danesch) - arid meadows - PC

E ***Ophrys promontorii*** O. Danesch & E. Danesch - arid meadows - PC

Fig. 8 *Ophrys exaltata* subsp. *archipelagi* (Photo by F. Conti)

Ophrys riojana C. E.Hermos. - arid meadows - R - Gioia dei Marsi-Pescasseroli, Ortona dei Marsi (Souche, in verbis).

E ***Ophrys tenthredinifera*** Willd. subsp. ***neglecta*** (Parl.) E. G. Camus (*O. neglecta* Parl.) - arid meadows - RR - loc. Querceto near Campoli Appennino (Minutillo in litt.).

Ophrys tetraloniae W. P. Teschner (*O. holosericea* (Burnm. f.) Greuter subsp. *tetraloniae* (W. P. Teschner) Kreutz; *O. serotina* H. Rolli ex Cortesi) - arid meadows, glades - R - near Alfedena (Reinhard 1987), loc. Nocicchia (Conti and Minutillo 1998), near Scontrone (Griebl 2010), Barrea (Souche, in verbis).

Orchis anthropophora (L.) All. (*Aceras anthropophorum* (L.) R. Br.) - arid meadows, spread from the low hill to about 1,000 m - C

Orchis italica Poir. - arid meadows, from the hill to about 1,200 m - PC

Orchis mascula (L.) L. subsp. ***mascula*** - cool pastures, glades - C

Orchis mascula (L.) L. subsp. ***speciosa*** (Mutel) Hegi (*O. mascula* (L.) L. subsp. *signifera* (Vest.) Soó; *O. signifera* Vest) - cool pastures, glades - C

Orchis militaris L. - glades, from 500 to about 1,600 m - PC

Orchis pallens L. - stony pastures from 1,000 to about 2,000 m - PC

Orchis pauciflora Ten. (*O. provincialis* Balbis subsp. *pauciflora* (Ten.) Camus) - arid meadows from the hill to about 1,500 m - C

Orchis provincialis Balb. ex Lam. and DC. - woods - R - Forca d'Acero (Falqui 1899); Camosciara near Rif. Liscia (Hennecke and Hennecke 1999).

Orchis purpurea Huds. (*O. fusca* Jacq.) - arid meadows and uncultivated land, from the hill to about 1,300 m - CC

Orchis simia Lam. - arid meadows, from the hill to about 1,200 m - PC

Orchis spitzelii Saut. ex W. D. J. Koch - subalpine scrub, pastures - PC

Orchis × colemanii Cortesi - margins of woods - PC

Platanthera algeriensis Batt. & Trab. - margins of woods - R - near Passo del Diavolo, near Barrea, near Alfedena (Griebl 2010).

Platanthera bifolia (L.) Rich. (*Orchis bifolia* L.) - open woods, meadows on a clayey substratum from 800 to about 1,400 m - PC

Platanthera chlorantha (Custer) Rchb. (*Orchis chlorantha* Custer) - margins of woods, glades from 600 to about 1,300 m - PC

Pseudorchis albida (L.) Á. Löve & D. Löve (*Gymnadenia albida* (L.) Rich.; *Leucorchis albida* (L.) E. Meyer; *Satyrium albidum* L.) - grassy and stony slopes subject to long periods of snow, often with *Dryas octopetala* - R - Camosciara, Forca Resuni, Picinisco allo Zaffineto (Terracciano 1878; Bazzichelli and Furnari 1970), Mainarde! (Conti 1992).

Serapias bergonii E. G. Camus (*S. vomeracea* (Burm. f.) Briq. subsp. *laxiflora* (Soó) Gölz & H. R. Reinhard) - arid meadows - RR - Scapoli near Ponte Nuovo (Hennecke and Hennecke 1999).

Serapias lingua L. - arid meadows - R - between Montenero and Alfedena!, Alfedena! (Conti and Pellegrini 1990; Conti 1998).

Serapias parviflora Parl. - arid meadows - PC

Serapias vomeracea (Burm. f.) Briq. (incl. *S. vomeracea* (Burm. f.) Briq. subsp. *longipetala* (Ten.) H. Baumann & Künkele) - arid meadows - PC

Spiranthes spiralis (L.) Chevall. (*Ophrys spiralis* L.; *S. autumnalis* Rich.) - arid meadows - PC

Traunsteinera globosa (L.) Rchb. - montane meadows - RR - Passo Godi (Conti and Pellegrini 1990 quote Daiss).

Orobanchaceae

Bellardia latifolia (L.) Cuatrec. (*Bartsia latifolia* (L.) Sibth. & Sm.; *Euphrasia latifolia* L.; *Parentucellia latifolia* (L.) Caruel) - arid meadows - C

Bellardia trixago (L.) All. (*Bartsia trixago* L.) - arid meadows - RR - near Castel S. Vincenzo! (Conti 1995).

Euphrasia alpina Lam. - D - Montagna di Frattura (Gravina 1812).

Euphrasia illyrica Wettst. - stony slopes - PC

Euphrasia italica Wettst. - arid pastures - PC

Euphrasia liburnica Wettst. - pastures - PC

Euphrasia micrantha Rchb. - D - the reports of this species (Tenore 1842; Tenore and Gussone 1842; Terracciano 1873) are probably erroneous.

Euphrasia officinalis L. subsp. ***rostkoviana*** (Hayne) Towns. (*E. rostkoviana* Hayne) - pastures - PC

Euphrasia officinalis L. subsp. ***kerneri*** (Wettst.) Eb. Fisch. (*E. kerneri* Wettst.; *E. picta* Wimmer subsp. *kerneri* (Wettst.) Yeo) - pastures RR - Lago di Castel S. Vincenzo! (Conti 1995).

Euphrasia salisburgensis Funck ex Hoppe - pastures, stony slopes - C

Euphrasia stricta D. Wolff ex J. F. Lehm. (*E. pectinata* Ten.) - arid pastures, open woods - C - the reports of *E. minima* are probably to be referred to this species.

Melampyrum arvense L. subsp. ***arvense*** - fields, uncultivated land - PC

E ***Melampyrum italicum*** Soó - open woods - R - S. Michele a Foce! (Conti 1995), Gole del Sagittario! (Conti and Tinti 2012), Valle Franchitta!

Melampyrum nemorosum L. - D - near Scanno (Anzalone and Bazzichelli 1960).

E ***Melampyrum variegatum*** Huter, Porta & Rigo - arid pastures - RR - Villavallelonga (Anzalone and Bazzichelli 1960).

Odontites luteus (L.) Clairv. (*Euphrasia lutea* L.) - arid meadows - CC

Odontites vernus (Bellardi) Dumort. (*Euphrasia verna* Bellardi) - meadows - PC

Odontites vulgaris Moench subsp. ***vulgaris*** (*O. rubra* (Baumg.) Opiz subsp. *rubra*) - meadows - PC

Orobanche alba Stephan ex Willd. - pastures, on *Thymus* - C

Orobanche amethystea Thuill. - arid pastures, on *Lotus* - R - Serra della Terratta, Serra della Cappella (Petriccione 1988)

Orobanche caryophyllacea Sm. - on *Rubiaceae* - PC

Orobanche crenata Forssk. - uncultivated land, fields, on *Leguminosae* - PC

E ***Orobanche ebuli*** Huter & Rigo - on *Sambucus ebulus* - R - Opi - Forca d'Acero! (Lattanzi et al. 2000), Prati d'Angro, Rif. Forca Resuni (Corazzi et al. 2003).

Orobanche elatior Sutton (*Orobanche major* L., nom. rej.) - D - alla Pietrosa, Monte Meta (Terracciano 1873).

Orobanche gracilis Sm. - pastures - C

Orobanche hederae Duby - woods - R - Gole del Sagittario! (Conti and Tinti 2012).

Orobanche minor Sm. - on *Leguminosae*, *Galium*, *Daucus* - PC

*****Orobanche nana*** (Reut.) Beck (*O. ramosa* L. subsp. *nana* (Reut.) Cout.; *Phelipanche nana* (Reut.) Soják) - fields, arid uncultivated land - RR - Monte Castelnuovo!

Orobanche purpurea Jacq. (*Phelipanche purpurea* (Jacq.) Soják) - arid meadows - R - Gole del Sagittario! (Conti and Minutillo 2001), below Castelnuovo!

Orobanche ramosa L. (*Phelipanche ramosa* (L.) Pomel) - fields, arid uncultivated land - PC

****Orobanche rapum-genistae*** Thuill. - pastures - RR - S. Michele a Foce near the Eremo! New species for Molise region.

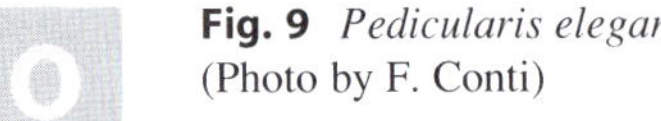

Fig. 9 *Pedicularis elegans* (Photo by F. Conti)

Orobanche reticulata Wallr. (*O. reticulata* Wallr. subsp. *pallidiflora* (Wimm. & Grab.) Hayek) - pastures - PC

*****Orobanche teucrii*** Holandre - pastures - R - Monte Castelnuovo!, Pizzone - Casone del medico!, Colle dell'Olmo di Bobbi near the tunnel of Carrito! New species for Molise region.

Pedicularis comosa L. subsp. ***comosa*** - stony - grassy slopes - CC

E ***Pedicularis elegans*** Ten. (*P. elegans* Ten. subsp. *praetutiana* (Levier ex Steininger) Pign.Wik.; *P. gyroflexa* Vill. subsp. *praetutiana* (Levier ex Steininger) H. Kunz & E. Mayer) (Fig. 9) - stony - grassy slopes - C

Pedicularis friderici-augusti Tomm. - stony - grassy slopes - R - Villavallelonga at Tre Solchi (Grande 1910), Rifugio Forca Resuni (Anzalone and Bazzichelli 1960).

Pedicularis hoermanniana K. Malý - scrub, *Fagus sylvatica* wood glades - PC

Pedicularis rostratospicata Crantz (Fig. 6 in chapter "Vegetation Features" and Fig. 10) - stony - grassy slopes, screes - R - Monte Petroso! (Anzalone and Bazzichelli 1960), Serra delle Gravare! (Bruno and Bazzichelli 1966), Monte Marsicano! (Rovelli and Conti 1995). This taxon deserves further study, actually does not fit in the known subspecies.

Pedicularis verticillata L. subsp. *verticillata* - D - generically recorded for Marsica (Lastoria 2000).

Rhinanthus alectorolophus (Scop.) Pollich subsp. ***alectorolophus*** - pastures, stony slopes - C

Rhinanthus minor L. (*R. personatus* (Behrend.) Bég.) - cool meadows - C

E ***Rhinanthus wettsteinii*** (Sterneck) Soó (*Alectorolophus wettsteinii* Sterneck) - stony pastures - C

Fig. 10 *Pedicularis rostratospicata* (Photo by F. Conti)

Oxalidaceae

Oxalis acetosella L. - *Fagus sylvatica* woods - PC

Oxalis corniculata L. (*O. repens* Thunb.; *O. tropaeoloides* Schlachter ex Planch.) - NC - Mainarde (Zodda 1931).

P

Paeoniaceae

E ***Paeonia officinalis*** L. subsp. ***italica*** N. G. Passal. & Bernardo (Fig. 1) - *Fagus sylvatica* wood glades, cool pastures - R - Scanno alla Fontana della Fascia (Gravina 1812), near Passo Godi! (D'Andrea 1982), Serra della Terratta! (Conti 1998).

Papaveraceae

Chelidonium majus L. - ruderal environments, walls - C

Corydalis cava (L.) Schweigg. & Körte subsp. ***cava*** - woods - C

Corydalis densiflora C. Presl (*C. solida* (L.) Clairv. subsp. *densiflora* (C. Presl) Arcang.) - margins of woods, open woods - R - Villavallelonga (Anzalone and Bazzichelli 1960), Serra Lunga!, Val Cervara! The populations of the central Apennines are currently under study by the authors.

Corydalis pumila (Host) Rchb. (*Fumaria pumila* Host) - woods - PC

Fumaria capreolata L. subsp. ***capreolata*** - fields, uncultivated land - PC

Fumaria officinalis L. subsp. ***officinalis*** - fields, uncultivated land, ruderal environments - C

Fumaria parviflora Lam. - fields - RR - Pescasseroli (Anzalone and Bazzichelli 1960).

Fumaria petteri Rchb. subsp. ***petteri*** - arid meadows - RR - Monte Castelnuovo! (Conti 1995).

F. Conti, F. Bartolucci, *The Vascular Flora of the National Park of Abruzzo, Lazio and Molise (Central Italy)*, Geobotany Studies, DOI 10.1007/978-3-319-09701-5_20

Fig. 1 *Paeonia officinalis* subsp. *italica* (Photo by F. Conti)

Papaver alpinum L. subsp. ***alpinum*** (*P. degenii* (Urum. & Jáv.) Kuzmanov; *P. ernesti-mayeri* (Markgr.) T. Wraber; *P. julicum* E. Mayer & Merxm.; *P. alpinum* L. subsp. *ernesti-mayeri* Markgr.) (Fig. 5 in chapter "Vegetation Features") - calcareous screes at high altitude - RR - Monte Marsicano! (Petriccione 1988).

****Papaver apulum*** Ten. - fields, uncultivated arid land - R - Gole del Sagittario! (Conti and Tinti 2012 sub *P. hybridum* L.), Collelongo!

Papaver dubium L. subsp. ***dubium*** - fields, uncultivated arid land - R - "S. Biagio Saracinisco nel salire alla Monna" (Terracciano 1873), Pizzo Marcello!, Templo!

Papaver rhoeas L. subsp. ***rhoeas*** (*P. rhoeas* L. var. *strigosum* Boenn.; *P. strigosum* (Boenn.) Schur) - fields - C

A ***Papaver somniferum*** L. - ruderal environments - CAS

Pseudofumaria alba (Mill.) Lidén subsp. ***alba*** (*Corydalis ochroleuca* Koch; *Fumaria alba* Mill.) - stony slopes and shady screes at the feet of the mountains - PC

Pinaceae

Pinus mugo Turra subsp. ***mugo*** - scrub and rocky slopes in the subalpine belt - R - Camosciara! (Terracciano 1873 sub *P. montana*; Fiori 1927 sub *P. pumilio*; Fiori and Béguinot 1927; Anzalone and Bazzichelli 1960 sub *P. mugo* var. *pumilio*; Bruno and Bazzichelli 1966; Stanisci 1994, 1997), V.ne della Terratta!

E ***Pinus nigra*** J. F. Arnold subsp. ***nigra*** var. ***italica*** Hochst. (*P. laricio* Poir. subsp. *poiretiana* (Antoine) Endl.var. *magellensis* Endl.) - montane slopes in woods and rocky places - R - Villetta Barrea, Civitella Alfedena, Camosciara, Valle di Cacciagrande, Tre Confini, Monte Marrone, Valle di Mezzo, ridge between Valle Ura and Valle Fredda, over the stazzo dell'Affogato (Terracciano 1872; Furrer 1928; Lusina 1954; Anzalone and Bazzichelli 1960; Conti 1995).

Plantaginaceae

A ***Antirrhinum majus*** L. subsp. ***majus*** - uncultivated land, walls - NAT

Callitriche brutia Petagna (*C. capillaris* Parl.; incl. *C. hamulata* Kütz. ex W. D. J. Koch; incl. *C. brutia* Petagna var. *hamulata* (Kütz. ex W. D. J. Koch) Lansdown) - sluggish waters - R - Bosco Frascaro on the right bank of Rio Chiaro near the confluence with F. Volturno (Conti and Minutillo 2001), Montenero Val Cocchiara!

Callitriche lenisulca* Clavaud - sluggish waters - RR - Montenero Val Cocchiara! New species for Molise region.

Callitriche platycarpa* Kütz. - sluggish waters - R - Montenero val Cocchiara!, Source of Volturno! New species for Molise region.

Callitriche stagnalis Scop. - sluggish or stagnant waters - RR - Sorgenti Cavuto! (Conti and Tinti 2012).

Chaenorhinum minus (L.) Lange subsp. ***minus*** - arid uncultivated land - C

E ***Cymbalaria glutinosa*** Bigazzi & Raffaelli subsp. ***glutinosa*** - cliffs - R - Lecce nei Marsi (Anzalone and Bazzichelli 1960), near Castelnuovo a Volturno!

Cymbalaria muralis G. Gaertn., B. Mey. & Scherb. subsp. ***muralis*** - walls and cliffs - PC

Cymbalaria muralis G. Gaertn., B. Mey. & Scherb subsp. ***visianii*** (Jav.) D. A. Webb - walls and cliffs - PC

E ***Cymbalaria pallida*** (Ten.) Wettst. (*Linaria pallida* Ten.) (Fig. 2) - screes - PC

Digitalis ferruginea L. - glades, open woods - C

E ***Digitalis micrantha*** Roth ex Schweigg. (*D. australis* Ten.; *D. lutea* L. subsp. *australis* (Ten.) Arcang.; *D. lutea* L. var. *micrantha* (Roth ex Schweigg.) Lindl.) - glades, woods - CC

Globularia bisnagarica L. (*G. punctata* Lapeyr.) - arid pastures - C

Globularia meridionalis (Podp.) O. Schwarz - montane stony slopes - CC

Kickxia commutata (Bernh. ex Rchb.) Fritsch subsp. *commutata* - observed at Rio Mollarino! (Conti and Minutillo 1998) outside the Park but close to the buffer external zone.

Fig. 2 *Cymbalaria pallida* (Photo by F. Conti)

Kickxia elatine (L.) Dumort. (*Antirrhinum elatine* L.) - NC - "Valvori nell'andare a S. Biagio Saracinisco" (Terracciano 1873), near S. Biagio (Zodda 1931).

Kickxia spuria (L.) Dumort. subsp. ***integrifolia*** (Brot.) R. Fern. - uncultivated land - R - Lago Vivo!, Pantanello!

Kickxia spuria (L.) Dumort. subsp. ***spuria*** - uncultivated land - RR - Gole del Sagittario! (Conti and Tinti 2012).

E ***Linaria purpurea*** (L.) Mill. - margins of woods, stony areas - CC

****Linaria simplex*** (Willd.) DC. (*Antirrhinum simplex* Willd.) - arid uncultivated areas, pastures, fields - RR - near Cocullo!

Linaria vulgaris Mill. subsp. ***vulgaris*** - roadsides, uncultivated land - C

Misopates orontium (L.) Raf. (*Antirrhinum orontium* L.) - NC - Picinisco (Tenore and Gussone 1842). Observed at Colle Truscino! outside the Park but close to the buffer external zone.

Plantago afra L. subsp. ***afra*** - arid uncultivated land - RR - between Lagone and Mastrogiovanni! (Conti and Minutillo 2001 sub *P. arenaria*).

Plantago argentea Chaix subsp. ***argentea*** - stony pastures - PC

Plantago atrata Hoppe subsp. ***atrata*** - meadows at high altitude, snowbed meadows - C

Plantago atrata Hoppe subsp. ***fuscescens*** (Jord.) Pilg. (*P. fuscescens* Jord.) - stony slopes - PC

Plantago coronopus L. - NC - Mainarde (Zodda 1931).

Plantago holosteum Scop. - arid pastures - C

Plantago lagopus L. - arid uncultivated land - RR - between Lagone and Mastrogiovanni (Conti and Minutillo 2001).

Plantago lanceolata L. - ruderal environments, uncultivated land - C

Plantago major L. subsp. ***major*** - uncultivated land, ruderal environments, plains at high altitude - PC - the populations at high altitude have small leaves 3–5 nerves (as in *P. major* subsp. *pleiosperma* and *P. major* subsp. *winteri*) but have few seeds, generally nine (as in *P. major* ss and *P. major* subsp. *winteri*). The subsp. *winteri* is probably to be excluded because it seems to be limited to salty depressions.

Plantago major L. subsp. ***sinuata*** (Lam.) Negodi - uncultivated land - R - Villavallelonga, Valle di Canneto (Anzalone and Bazzichelli 1960).

Plantago maritima L. subsp. ***serpentina*** (All.) Arcang. (*P. serpentina* All.) - high pastures - R - La Meta (Rossi and Bassani 1982), Alta Valle Pagana!, Rocca Altiera.

Plantago media L. subsp. ***media*** - meadows and pastures - C

Plantago sempervirens Crantz (*Plantago cynops* L.) - arid meadows - C

Veronica acinifolia L. - fields - RR - Villavallelonga (Anzalone and Bazzichelli 1960).

Veronica agrestis L. - fields, sheepfolds - R - Monte S. Marcello near S. Donato (Falqui 1899), Prati di Mezzo!, Fontana Cecalupo near Settefrati (Conti 1995)

Veronica alpina L. (*V. pumila* All.) - pastures - R - Zaffineto, Tre Confini, Tabaccaro (Terracciano 1873), Monte Palombo (Petriccione 1986).

Veronica anagallis-aquatica L. subsp. ***anagallis-aquatica*** - ditches, riversides - C

Veronica aphylla L. subsp. ***aphylla*** - humid cliffs, stony slopes at high altitude - C
Veronica arvensis L. - arid meadows, fields, ruderal environments - CC
Veronica barrelieri H. Schott ex Roem. & Schult. subsp. ***barrelieri*** (*Pseudolysimachion barrelieri* (H. Schott ex Roem. & Schult.) Holub subsp. *barrelieri*) - pastures - RR - Serra Lunga between Civita d'Antino and Collelongo! (Conti 1998; Conti and Minutillo 1998).
Veronica beccabunga L. - ditches, riversides - CC
Veronica chamaedrys L. subsp. ***chamaedrys*** - glades, scrub - CC
Veronica cymbalaria Bodard subsp. ***cymbalaria*** - stony slopes, ruderal environments - PC
Veronica fruticans Jacq. - NC - Monte Meta (Terracciano 1872).
Veronica hederifolia L. subsp. ***hederifolia*** - fields, open woods - C
Veronica montana L. - cool woods - C
Veronica officinalis L. - open woods, pastures - C
Veronica orsiniana Ten. subsp. ***orsiniana*** - arid pastures, glades - C
A ***Veronica persica*** Poir. - ruderal environments - INV
Veronica polita Fr. - fields, ruderal environments - PC
Veronica praecox All. - stony slopes - R - Forca d'Acero (Terracciano 1878), Gole del Sagittario! (Conti and Tinti 2012).
Veronica prostrata L. subsp. ***prostrata*** - arid and stony pastures - PC
Veronica scutellata L. - humid meadows - R - Le Forme! (Conti 1992), near Opi, Campitelli! (Buchwald 1995; Conti 1998).
Veronica serpyllifolia L. (incl. *V. apennina* Tausch; incl. *V. serpyllifolia* L. subsp. *apennina* (Tausch) Rouy; incl. *V. serpyllifolia* L. subsp. *humifusa* (Dicks.) Syme; incl. *V. humifusa* Dicks.; incl. *V. serpyllifolia* L. var. *humifusa* (Dickson) Sm.) - glades and cool meadows - C
Veronica verna L. subsp. ***verna*** - arid pastures - R - Colle Biferno!, Le Forme! (Conti 1998; Conti and Minutillo 1998).

Platanaceae

A ***Platanus orientalis*** L. - ruderal environments - NAT - Collelongo (Viegi et al. 1990).

Plumbaginaceae

Armeria arenaria (Pers.) Schult. in Roem. & Schult. subsp. *arenaria* (*A. plantaginea* Willd.; *A. alliacea* auct. Fl. Ital.; *Statice arenaria* Pers.; *St. plantaginea* All.) - D - Villavallelonga, Barrea (Vaccari and Wilczek 1940; Anzalone and Bazzichelli 1960).
Armeria canescens (Host) Ebel (*A. gracilis* Ten.; *Statice canescens* Host; *A. majellensis* Boiss.; *A. majellensis* Boiss. subsp. *ausonia* Bianchini;

A. canescens (Host) Ebel subsp. *gracilis* (Ten.) Bianchini) - montane stony pastures - CC

Plumbago europaea L. - ruderal environments, uncultivated land - PC

Poaceae

Agrostis capillaris L. subsp. ***capillaris*** (*A. tenuis* Sibth.) - pastures - PC

Agrostis castellana Boiss. & Reut. (incl. *A. parlatorei* Breistr.) - glades, humid uncultivated land - R - Picinisco alla Forestella (Terracciano 1890), Pantanello of Montenero Val Cocchiara!, Valle Inguagnera!

Agrostis gigantea Roth subsp. ***gigantea*** - humid environments - RR - Monte Ara dei Merli (Petriccione et al. 1994).

*****Agrostis stolonifera*** L. subsp. ***scabriglumis*** (Boiss. & Reut.) Maire (*A. scabriglumis* Boiss. & Reut.) - humid meadows - RR - Lago Vivo!

Agrostis stolonifera L. subsp. ***stolonifera*** (*A. stolonifera* L. subsp. *stolonifera* var. *densiflora* (Guss.) Giardina & Raimondo) - humid meadows - C

Aira caryophyllea L. subsp. ***caryophyllea*** - arid meadows - R - Picinisco (Tenore and Gussone 1842), Collicillo of Villavallelonga (Conti 1995 from a specimen collected by Rosati).

Alopecurus aequalis Sobol. - marshy environments - PC

Alopecurus alpinus Vill. (*A. gerardi* Vill.) - snowbed meadows - PC

Alopecurus myosuroides Huds. subsp. ***myosuroides*** - uncultivated land, fields - PC

Alopecurus pratensis L. subsp. ***pratensis*** - humid meadows - PC

Alopecurus rendlei Eig (*A. utriculatus* auct., non Sol.) (Fig. 3) - humid meadows - PC

Ampelodesmos mauritanicus (Poir.) T. Durand & Schinz (*A. tenax* (Vahl) Link; *Arundo mauritanica* Poir.) - garigue - R - near Picinisco (Spada 1979), near the Abbazia of Castel S. Vincenzo!, between Collalto and Scapoli! (Conti 1995), Vallone Lacerno!

Anisantha diandra (Roth) Tutin (*Bromus ambigens* Jord.; *B. rigidus* Roth subsp. *ambigens* (Jord.) Pignatti; *B. diandrus* Roth; *B. gussonei* Parl.) - arid uncultivated land - PC

Anisantha madritensis (L.) Nevski subsp. ***madritensis*** (*Bromus madritensis* L. subsp. *madritensis*) - arid uncultivated land, ruins - PC

Anisantha rigida (Roth) Hyl. (*Bromus rigidus* Roth; *B. diandrus* Roth subsp. *maximus* (Desf.) Soó) - arid uncultivated land - PC

Anisantha sterilis (L.) Nevski (*Bromus sterilis* L.) - arid uncultivated land - CC

Anisantha tectorum (L.) Nevski (*Bromus tectorum* L.) - arid uncultivated land - C

Anthoxanthum odoratum L. - meadows - C - some individuals could be referred to subsp. *nipponicum* (Honda) Tzvelev rather than the subsp. *odoratum*, however, the differences between these two *taxa* are not clear.

Aristella bromoides (L.) Bertol. (*Agrostis bromoides*L.; *Achnatherum bromoides* (L.) P. Beauv.; *Lasiagrostis bromoides*(L.) Nevski; *Stipa bromoides*(L.) Dorfl.) - maquis, garigue - R - Vallone Lacerno! Also recorded near Lecce nei Marsi and

Fig. 3 *Alopecurus rendlei* (Photo by F. Conti)

Casale di Aschi (Anzalone and Bazzichelli 1960) outside the Park but close to the buffer external zone.

Arrhenatherum elatius (L.) P. Beauv. ex J. Presl & C. Presl subsp. ***elatius*** - meadows - C

Arundo collina Ten. - clayey slopes - RR - S.ta Lucia near Castelnuovo a Volturno! (Conti 1995).

Avena barbata Pott ex Link subsp. ***barbata*** - meadows and uncultivated land - R - Gole del Sagittario! (Conti and Tinti 2012) and probably more widespread.

Avena fatua L. subsp. ***fatua*** - arid uncultivated land, fields - R - S. Michele a Foce! (Conti 1995), Gole del Sagittario! (Conti and Tinti 2012).

Avena sterilis L. - meadows, fields - RR - Valle di Canneto (Anzalone and Bazzichelli 1960). The infraspecific rank is not indicated by the authors.

Avenula pubescens (Huds.) Dumort. subsp. ***pubescens*** (*Avena pubescens* Huds.; *Avenastrum pubescens* Jessen; *Homalotrichon pubescens* (Huds.) Banfi, Galasso & Bracchi, *nom. illeg.*; *Helictotrichon pubescens* (Huds.) Pilg.) - arid meadows - PC

Bellardiochloa variegata (Lam.) Kerguélen subsp. ***variegata*** - montane pastures and meadows at high altitude - PC

Bothriochloa ischaemum (L.) Keng (*Andropogon ischaemum* L.; *Dichanthium ischaemum* (L.) Roberty) - arid meadows - PC

P

E ***Brachypodium genuense*** (DC.) Roem. & Schult. - subalpine pastures - C

Brachypodium rupestre (Host) Roem. & Schult. (*B. pinnatum* (L.) Beauv. subsp. *rupestre* (Host) Schübl. & G. Martens ; *Bromus rupestris* Host) - arid pastures - CC

Brachypodium sylvaticum (Huds.) P. Beauv. - open woods - CC

Briza maxima L. - fields, arid uncultivated land - PC

Briza media L. - meadows, uncultivated land - PC

Briza minor L. - uncultivated land, meadows - PC

Bromopsis benekenii (Lange) Holub (*Bromus benekenii* (Lange) Trimen) - woods - PC

Bromopsis caprina (A. Kern. ex Hack.) Banfi & N. G. Passal. - arid meadows - R - generically recorded for Marsica (Pignatti 1982).

Bromopsis erecta (Huds.) Fourr. subsp. ***erecta*** (*Bromus erectus* Huds. subsp. *erectus*) - meadows and arid pastures - CC

Bromopsis inermis (Leyss.) Holub (*Bromus inermis* Leyss.) - borders of the roads - RR - Val di Sangro near the Casone Crugnale! (Conti and Minutillo 2001).

****Bromopsis pannonica*** (Kumm. & Sendtn.) Holub subsp. ***pannonica***(*Bromus pannonicus* Kumm. & Sendtn.) - woods - RR - Monte Mezzana!, over Liscia! New taxon for Abruzzo region.

Bromopsis ramosa (Huds.) Holub subsp. ***ramosa*** (*Bromus ramosus* Huds. subsp. *ramosus*) - woods - C

Bromus arvensis L. subsp. ***arvensis*** - fields, ruins - R - Barrea (Tenore and Gussone 1842), Lago Vivo (Anzalone and Bazzichelli 1960).

Bromus commutatus Schrad. subsp. ***commutatus*** - uncultivated land - R - Macchiarvana (Anzalone and Bazzichelli 1960), Fonte Vetica!, Fonte Campagliona! (Conti 1995).

Bromus hordeaceus L. subsp. ***molliformis*** (J. Lloyd ex Billot) Maire & Weiller (*B. hordeaceus* L. subsp. *divaricatus* auct. Fl. Ital.) - garigue - C - the reports of *B. hordeaceus* subsp. *hordeaceus* for Mainarde (Conti 1995) are to be referred to this taxon.

Bromus intermedius Guss. (*B. intermedius* Guss. subsp. *divaricatus* Bonnier & Layens) - arid pastures - R - Morrone delle Rose, Monte Mattone (Conti 1995).

Bromus japonicus Thunb. subsp. ***japonicus*** - uncultivated land - R - Gioia Vecchio, Villetta Barrea (Anzalone and Bazzichelli 1960).

Bromus racemosus L. subsp. ***racemosus*** - meadows - PC

Bromus secalinus L. - fields - RR - Cocullo (Greco and Petriccione 1989).

Bromus squarrosus L. subsp. ***squarrosus*** - uncultivated land, arid meadows - C

Calamagrostis varia (Schrad.) Host (*Arundo varia* Schrad.; *C. montana* Host) - open woods, glades - C

Catabrosa aquatica (L.) P. Beauv. - humid meadows - RR - Piana di Pescasseroli (Pedrotti et al. 1992).

Catapodium rigidum (L.) C. E. Hubb. subsp. ***rigidum*** (*Desmazeria rigida* (L.) Tutin subsp. *rigida*) - arid meadows and uncultivated land - C

*A ***Ceratochloa cathartica*** (Vahl) Herter (*Bromus catharticus* Vahl; *Bromus willdenowii* Kunth; *Ceratochloa willdenowii* (Kunth) W. A. Weber) - uncultivated land - NAT - Val Fondillo sources near the Segheria!

****Cleistogenes serotina*** (L.) Keng subsp. ***serotina***(*Kengia serotina* (L.) Packer; *Diplachne serotina* (L.) Link; *Festuca serotina* L.) - stony and arid meadows - R - Vallone Lacerno!, near Settefrati!

Crypsis alopecuroides (Piller & Mitterp.) Schrad. (*C. nigricans* Guss.; *Heleochloa alopecuroides* (Piller & Mitterp.) Host ex Roem.; *Phleum alopecuroides* Piller & Mitterp.) - lake margins - RR - Lago di Barrea! (Conti et al. 2011b).

Cynodon dactylon (L.) Pers. - ruderal environments, uncultivated land - PC

Cynosurus cristatus L. (*C. polybracteatus* Poir.) - meadows - C

Cynosurus echinatus L. - arid meadows - C

Cynosurus effusus Link (*C. elegans* auct., non Desf.) - arid uncultivated land - R - Villetta Barrea, Collelongo, Val Canneto (Anzalone and Bazzichelli 1960).

Dactylis glomerata L. subsp. ***glomerata*** - uncultivated land, ruderal environments - C

Dactylis glomerata L. subsp. ***hispanica*** (Roth) Nyman (*D. hispanica* Roth) - uncultivated land and arid meadows - PC

Dactylis glomerata L. subsp. ***lobata*** (Drejer) H. Lindb. (*D. aschersoniana* Graebn.; *D. glomerata* L. subsp. *aschersoniana* (Graebn.) Thell.; *D. polygama* Horv.) - pastures - RR - Cava di Rena! (Conti and Bartolucci 2011a).

Danthonia decumbens (L.) DC. subsp. ***decumbens*** (*Festuca decumbens* L.; *Sieglingia decumbens* (L.) Bernh.) - humid pastures, meadows at high altitude - R - Val Fondillo!, Camosciara! (Conti and Minutillo 2001), Monte Genzana!

Dasypyrum villosum (L.) P. Candargy (*Agropyron villosum* (L.) Link; *Haynaldia villosa* (L.) Schur; *Pseudosecale villosum* (L.) Degen; *Secale villosum* L.; *Triticum villosum* (L.) M. Bieb.) - arid uncultivated areas and pastures - PC

Deschampsia cespitosa (L.) P. Beauv. subsp. ***cespitosa*** - humid meadows - CC

Digitaria ischaemum (Schreb. ex Schweigg.) Schreb. ex Muhl. subsp. *ischaemum* - ruderal environments, riversides - NC - Picinisco (Tenore and Gussone 1842 sub *D. humifusa*). Also observed at Rio Mollarino! (Conti and Minutillo 1998) outside the Park but close to the buffer external zone.

Digitaria sanguinalis (L.) Scop. subsp. *sanguinalis* - observed at Rio Mollarino! (Conti and Minutillo 1998) outside the Park but close to the buffer external zone.

*E ***Drymochloa drymeja*** (Mert. & W. D. J. Koch) Holub subsp. ***exaltata*** (C. Presl) Foggi & Signorini (*Festuca drymeja* auct. Fl. Ital.; *Festuca exaltata* C. Presl) - open woods - R - M. Falconara!

Drymochloa sylvatica (Pollich) Holub (*Festuca altissima* All.) - *Fagus sylvatica* woods - PC

Echinaria capitata (L.) Desf. (*Cenchrus capitatus* L.) - arid meadows - PC

Echinochloa crusgalli (L.) P. Beauv. subsp. ***crusgalli*** - cultivated areas, humid uncultivated land, ruderal environments - PC

A ***Eleusine tristachya*** (Lam.) Lam. - arid uncultivated area - CAS - near S. Donato Val di Comino (Conti and Minutillo 2001), near Campoli Appennino!

Elymus caninus (L.) L. (*Agropyron caninum* (L.) P. Beauv.) - open woods - C

Elytrigia repens (L.) Nevski subsp. ***repens*** (*Agropyron repens* (L.) P. Beauv.; *Triticum repens* L.; *Elymus repens* (L.) Gould subsp. *repens*) - uncultivated land, arid meadows - C

E ***Festuca alfrediana*** Foggi & Signorini subsp. ***ferrariniana*** Foggi, Parolo & Graz. Rossi (*F. vizzavonae* auct. p.p.) - stony slopes, cliffs at high altitude - PC

Festuca bosniaca Kumm. & Sendtn. subsp. ***bosniaca*** - stony slopes - PC

Festuca circummediterranea Patzke - stony slopes - C

Festuca heterophylla Lam. - woods - PC

Festuca inops De Not. (*F. gracilior* (Hack.) Markgr. - Dann.) - thermophilous stony pastures - C

Festuca jeanpertii (St. - Yves) Markgr. - Dann. subsp. ***campana*** (N. Terracc.) Markgr. - Dann. (*F. campana* (N. Terracc.) E. B. Alexeev) - stony slopes - PC

Festuca laevigata Gaudin (*F. curvula* Gaudin; incl. *F. laevigata* Gaudin subsp. *crassifolia* (Gaudin) Kerguélen & Plonka; incl. *F. crassifolia* (Gaudin) Landolt; incl. *F. glauca* Vill. var. *crassifolia* Gaudin) - stony pastures - C

Festuca microphylla (St. - Yves) Patzke (*F. nigrescens* Lam. subsp. *microphylla* (St. - Yves in Coste) Markgr. - Dann.; *F. nigrescens* Lam. subsp. *microphylla* Foggi & Graz. Rossi ; *F. rubra* L. subsp. *microphylla* St. - Yves in Coste) - snowbed meadows - PC

Festuca nigrescens Lam. (*F. rubra* L. subsp. *commutata* (Gaudin) Markgr. - Dann.) - acidophilous pastures - C

Festuca rubra L. subsp. ***rubra*** - meadows and pastures - PC

Festuca stricta Host subsp. ***trachyphylla*** (Hack.) Patzke ex Pils (*F. brevipila* R. Tracey; *F. trachyphylla* (Hackel) Krajina) - arid pastures - RR - near the Rif. Forca Resuni (Anzalone and Bazzichelli 1960).

E ***Festuca violacea*** Ser. ex Gaudin subsp. ***italica*** Foggi, Gr. Rossi & Signorini (*F. macrathera* (Hackel) Markgr. - Dann.; *F. violacea* Schleicher ex Gaudin subsp. *macrathera* (Hackel ex G. Beck) Markgr. - Dannenb) - meadows at high altitude - PC

Glyceria fluitans (L.) R. Br. (*Festuca fluitans* L.) - NC - Picinisco (Tenore and Gussone 1842).

Glyceria notata Chevall. (*G. fluitans* (L.) R. Br. subsp. *plicata* Fr.; *G. plicata* (Fr.) Fr.) - ditches, watercourses, marshy environments - C

Hainardia cylindrica (Willd.) Greuter (*Lepturus cylindricus* (Willd.) Trin.; *Parapholis cylindrica* (Willd.) C. E. Hubb.; *Rottboellia cylindrica* Willd.) - NC - Picinisco (Tenore and Gussone 1842; Terracciano 1873).

E ***Helictochloa praetutiana*** (Parl. ex Arcang.) Bartolucci, F. Conti, Peruzzi & Banfi subsp. ***praetutiana*** - meadows at high altitude, stony slopes - CC

Helictochloa pratensis (L.) Romero Zarco subsp. *pratensis* (*Avena pratensis* L.; *Avenula pratensis* (L.) Dumort.) - NC - Monte Forcellone (Tenore and Gussone 1842), S. Donato on Monte Croce (Terracciano 1878), Pescasseroli (Montelucci 1958 from a specimen collected by Fiori).

Holcus lanatus L. subsp. ***lanatus*** - humid meadows - C

Holcus mollis L. subsp. ***mollis*** - uncultivated land - RR - Camosciara (Anzalone and Bazzichelli 1960).

Hordelymus europaeus (L.) Harz (*Elymus europaeus* L.; *Hordeum europaeum* (L.) All.) - *Fagus sylvatica* woods - PC

Hordeum bulbosum L. - uncultivated land - RR - Fonte Vetica! (Conti 1995).

Hordeum murinum L. subsp. ***leporinum*** (Link) Arcang. (*H. leporinum* Link) - ruderal environments, arid uncultivated land - C

Hordeum secalinum Schreb. - humid meadows - R - Il Pantano (Pedrotti 1983), Piana di Pescasseroli (Pedrotti et al. 1992), near Opi (Buchwald 1995).

A ***Hordeum vulgare*** L. (incl. *H. distichon* L.) - uncultivated land - CAS - Val Canneto (Anzalone and Bazzichelli 1960).

Hyparrhenia hirta (L.) Stapf subsp. ***hirta*** (*Andropogon hirtus* L.; *Cymbopogon hirtus* (L.) Janch.) - arid meadows, garigue - R - near Lago di Grotta Campanaro (Spada 1979), Vallone Lacerno!

E ***Koeleria splendens*** C. Presl (*K. grandiflora* Bertol.; incl. *K. splendens* C. Presl subsp. *grandiflora* (Bertol. ex Schult.) Domin; *K. lobata* auct. Fl. Ital.) - arid and stony pastures - CC

Leucopoa dimorpha (Guss.) H. Scholz & Foggi (*Festuca dimorpha* Guss.) - screes - C

Lolium multiflorum Lam. (*L. multiflorum* Lam. subsp. *gaudini* (Parl.) Schinz & R. Keller ; *L. gaudini* Parl.) - meadows, uncultivated land - PC

Lolium perenne L. - uncultivated land, ruderal environments - C

A ***Lolium remotum*** Schrank - fields, uncultivated land - CAS - near S. Biagio (Zodda 1931).

Lolium rigidum Gaudin subsp. *rigidum* (*Lolium strictum* C. Presl) - NC - Picinisco, Barrea (Tenore and Gussone 1842).

Lolium temulentum L. (incl. *L. gussonei* Parl.; incl. *L. temulentum* L. subsp. *gussonei* (Parl.) Arcang.) - fields - PC

Melica ciliata L. subsp. ***ciliata*** - arid meadows - PC

Melica ciliata L. subsp. ***glauca*** (F. W. Schultz) K. Richt. (*M. glauca* F. W. Schultz) - arid meadows - PC

Melica nutans L. - *Fagus sylvatica* woods - R - Camosciara!, Mainarde! (Conti 1995, 1998).

Melica transsilvanica Schur subsp. ***klokovii*** Tzvelev - arid meadows - R - Valle di Canneto (Anzalone and Bazzichelli 1960; Petriglia 2004), near Pizzone!, Morrone delle Rose! (Conti 1995). The previous reports of *M. transsilvanica* subsp. *transsilvanica* are to be referred to this taxon (see Hempel 2011).

Melica uniflora Retz. - woods - CC

Milium effusum L. subsp. ***effusum*** - woods, glades - R - Valle di Canneto (Tenore and Gussone 1842) and generically for *Fagus sylvatica* woods (Bruno and Bazzichelli 1966).

Nardurus unilateralis (L.) Boiss. (*Festuca maritima* L.; *N. maritimus* (L.) Murb.; *Triticum unilaterale* L.; *Vulpia hispanica* auct. Fl. Ital., *V. unilateralis* (L.) Boiss.) - arid meadows - RR - near Sperone! (Conti 1995).

Nardus stricta L. - pastures - C

Oloptum miliaceum (L.) Röser & Hamasha (*Piptatherum miliaceum* (L.) Coss.; *Oryzopsis miliacea* (L.) Asch. & Schweinf.) - open termophilous woods, maquis

- RR - Val Canneto (Petriglia 2004). The nomenclature according to Hamasha et al. (2012).

Oloptum thomasii (Duby) Banfi & Galasso (*Milium thomasii* Duby; *Oryzopsis miliacea* (L.) Asch. & Schweinf. subsp. *thomasii* (Duby) Pignatti; *P. thomasii* (Duby) Kunth; *P. miliaceum* (L.) Coss. subsp. *thomasii* (Duby) Freitag) - open termophilous woods, maquis - R - Colleruta! (Conti and Minutillo 1998), Monte Falconara! The nomenclature according to Banfi and Galasso (2014).

Panicum dichotomiflorum Michx. - Also observed at Rio Mollarino! (Conti and Minutillo 1998) outside the Park but close to the buffer external zone.

Patzkea paniculata (L.) G. H. Loos subsp. ***paniculata*** (*Festuca paniculata* (L.) Schinz & Thell. subsp. *paniculata*; *F. spadicea* auct.) - montane pastures - PC

Phalaris brachystachys Link - fields - RR - Casale near Picinisco! (Conti and Minutillo 1998).

Phalaroides arundinacea (L.) Rauschert subsp. ***arundinacea*** - humid meadows, riversides - PC

Phleum hirsutum Honck. subsp. ***ambiguum*** (Ten.) Tzvelev (*Ph. ambiguum* Ten.) - arid pastures - CC

Phleum nodosum L. (*Ph. pratense* L. subsp. *bertolonii* (DC.) Bornm.; *Ph. bertolonii* DC.; *Ph. pratense* L. subsp. *serotinum* (Jord.) Berher) - meadows - PC

Phleum paniculatum Huds. subsp. *paniculatum* - NC - Picinisco (Tenore and Gussone 1842).

Phleum pratense L. subsp. ***pratense*** - cool meadows - PC

Phleum rhaeticum (Humphries) Rauschert (*Ph. alpinum* L. subsp. *rhaeticum* Humphries) - pastures at high altitude - C

Phleum subulatum (Savi) Asch. & Graebn. subsp. ***subulatum*** (*Phalaris subulata* Savi) - uncultivated land, arid pastures - R - Villavallelonga!, Val Canneto (Anzalone and Bazzichelli 1960).

Phragmites australis (Cav.) Trin. subsp. ***australis*** - humid environments - PC

Piptatherum virescens (Trin.) Boiss. (*Oryzopsis virescens* (Trin.) Beck; *Urachne virescens* Trin.) - scrub, hedges - NC - territory of Ortucchio dei Marsi (Grande 1913).

Poa alpina L. subsp. ***alpina*** - montane pastures, meadows at high altitude, screes, stony slopes - CC

****Poa angustifolia*** L. (*P. pratensis* L. subsp. *angustifolia* (L.) Gaudin) - humid environments - R - Valle Sorgiara!, il Lagozzo!

Poa annua L. (incl. *P. annua* L. var. *pilantha* Ronniger) - uncultivated land, ruderal environments - C

Poa bulbosa L. subsp. ***bulbosa*** - arid meadows, uncultivated land - C

Poa compressa L. - humid uncultivated land - PC

Poa infirma Kunth - uncultivated land - PC

Poa laxa Haenke subsp. *laxa* - NC - generically recorded for the Parco (Sipari 1926 quotes Pirotta; Lusina 1954).

Poa molinerii Balb. - stony slopes - PC

Poa nemoralis L. subsp. ***nemoralis*** - woods - C

**Poa perconcinna* J. R. Edm. (*P. concinna* auct.) - arid pastures - R - Valle Cupella!, Stazzo il Prato - Monte Greco! Recently confirmed to Abruzzo from Gran Sasso (Conti and Bartolucci 2011b).

Poa pratensis L. subsp. ***pratensis*** - cool meadows - CC

Poa trivialis L. subsp. ***sylvicola*** (Guss.) H. Lindb. (*P. sylvicola* Guss.) - woods - PC

Poa trivialis L. subsp. ***trivialis*** - meadows - C

Psilurus incurvus (Gouan) Schinz & Thell. (*Nardus aristatus* L.; *N. incurva* Gouan; *Psilurus aristatus* (L.) Duval - Jouve; *P. nardoides* Trin.) - meadows - D - meadows of Chiarano over the Piano di Cinque Miglia, Monte Forcellone (Tenore 1835).

Rostraria cristata (L.) Tzvelev (*Festuca cristata* L.; *Lophochloa cristata* (L.) Hyl.) - arid and stony pastures - PC

Schedonorus arundinaceus (Schreb.) Dumort. subsp. ***fenas*** (Lag.) H. Scholz (*Festuca fenas* Lag.; *F. arundinacea* Schreb. subsp. *fenas* (Lag.) Arcang.) - fields, uncultivated land, humid environments - PC - the reports of *S. arundinaceus* subsp. *arundinaceus* (Conti 1995) are to be referred to this taxon.

Schedonorus giganteus (L.) Holub (*Festuca gigantea* (L.) Vill.) - humid woods - R - Camosciara (Anzalone and Bazzichelli 1960), valle del Melfa below Picinisco!

Schedonorus pratensis (Huds.) P. Beauv. subsp. *apenninus* (De Not.) H. Scholz & Valdés (*Festuca apennina* De Not.; *F. pratensis* Huds. subsp. *apennina* (De Not.) Hegi) - NC - Picinisco in valle della Melfa (Terracciano 1873).

A ***Secale cereale*** L. subsp. ***cereale*** - uncultivated land - CAS - Gole di Barrea (Anzalone and Bazzichelli 1960).

Secale strictum (C. Presl) C. Presl subsp. ***strictum*** (*S. montanum* Guss.) - uncultivated land - R - Valle Ciavolara! (Conti 1995), Valle di Chiarano!

Sesleria autumnalis (Scop.) F. W. Schultz (*S. tuzsonii* Ujhelyi) - open woods - PC

Sesleria juncifolia Suffren subsp. ***juncifolia*** (*S. tenuifolia* Schrad.) - stony slopes - CC

E ***Sesleria nitida*** Ten. (incl. *S. nitida* Ten. subsp. *aprutia* Brullo & Giusso) - stony pastures - CC

A ***Setaria italica*** (L.) P. Beauv. subsp. ***italica*** - fields - CAS - Valle di Canneto (Viegi et al. 1990).

A ***Setaria pumila*** (Poir.) Roem. & Schult. (*Panicum glaucum* L.; *Setaria glauca* (L.) Beauv.) - fields - NAT - Picinisco (Tenore and Gussone 1842), Mainarde (Zodda 1931), Val Canneto (Petriglia 2004).

Setaria verticillata (L.) P. Beauv. (*Panicum verticillatum* L.; *Setaria ambigua* Guss.) - NC - Picinisco (Tenore and Gussone 1842), Mainarde (Zodda 1931).

Setaria viridis (L.) P. Beauv. subsp. ***viridis*** (*Panicum viride* L.; *Setariopsis viridis* (L.) Samp.) - fields - R - Picinisco (Tenore and Gussone 1842), Val Canneto, Villavallelonga (Anzalone and Bazzichelli 1960).

A ***Sorghum halepense*** (L.) Pers. - arid meadows - NAT - Colleruta (Conti and Minutillo 1998).

A ***Sporobolus indicus*** (L.) R. Br. (*S. poiretii* (Roem. & Schult.) Hitchc.) - uncultivated land - NAT - near Settefrati! (Conti and Minutillo 1998), near Campoli Appennino!

Stipa capillata L. - arid and stony slopes - PC

E ***Stipa dasyvaginata*** Martinovský subsp. ***apenninicola*** Martinovský & Moraldo - arid and stony slopes - C

Trachynia distachya (L.) Link (*Brachypodium distachyon* (L.) P. Beauv.; *Bromus distachyos* L.) - arid meadows - PC

Trisetaria flavescens (L.) Baumg. subsp. ***flavescens*** (*Trisetum flavescens* (L.) Beauv. subsp. *flavescens*) - cool woods - R - Piana di Pescasseroli (Pedrotti et al. 1992), Passo Godi!

Trisetaria panicea (Lam.) Paunero (*Avena panicea* Lam.; *Trisetum paniceum* (Lam.) Pers.) - uncultivated land - RR - Tre Ponti inferiore near S. Donato (Conti and Minutillo 2001).

E ***Trisetaria villosa*** (Bertol.) Banfi & Soldano (*Avena villosa* Bertol.; *Trisetum bertolonii* Jonsell) - cliffs - R - Monte Meta (Terracciano 1890; Bazzichelli and Furnari 1970), Gole del Sagittario (Pirone et al. 1997), Camosciara! (Conti 1998).

A ***Triticum aestivum*** L. (incl. *T. spelta* L.) - uncultivated land - CAS

Triticum ovatum (L.) Raspail (*T. ovatum* (L.) Raspail; *Aegilops geniculata* Roth) - arid meadows - PC

Vulpia ciliata Dumort. subsp. ***ciliata*** (*Festuca barbata* Gaudin; *F. ciliata* DC.) - arid meadows - PC

Vulpia myuros (L.) C. C. Gmel. subsp. ***myuros*** (*Festuca myuros* L. subsp. *myuros*) - arid uncultivated land, garigue - R - near Picinisco! (Terracciano 1890), near Scontrone! (Conti and Minutillo 2001).

A *Zea mays* L. - CAS - Mainarde (Zodda 1931).

Polygalaceae

Polygala alpestris Rchb. (incl. *P. alpestris* Rchb. subsp. *angelisii* (Ten.) Nyman; incl. *P. angelisii* Ten.) - montane pastures, stony slopes C

Polygala chamaebuxus L. - *Pinus nigra* woods, stony slopes - R - Camosciara!, Villetta Barrea!, Val Canneto (Tenore 1835; Tenore and Gussone 1842; Terracciano 1873, 1878; Anzalone and Bazzichelli 1960; Bruno and Bazzichelli 1966; Bazzichelli and Furnari 1970; Stanisci 1997), Mainarde! (Conti et al. 1990), Marsicano! (Conti 1995).

E ***Polygala flavescens*** DC. - arid pastures, margins of thermophilous woods - PC

Polygala major Jacq. - montane stony pastures - C

Polygala monspeliaca L. - garigue, arid meadows - R - near the Abbazia of Castel S. Vincenzo! (Conti 1995), Marsica! (Conti 1998).

Polygala nicaeensis W. D. J. Koch subsp. ***mediterranea*** Chodat - arid meadows - CC

Polygala vulgaris L. subsp. *vulgaris* - NC - generically recorded for the park (Sipari 1926 quotes Pirotta; Rovesti and Rovesti 1934).

Polygonaceae

Bistorta officinalis Delarbre (*Persicaria bistorta* (L.) Samp.; *Polygonum bistorta* L.) - cool montane meadows - PC

Bistorta vivipara (L.) Delarbre (*Persicaria vivipara* (L.) Ronse Decr.; *Polygonum viviparum* L.) - terraces and high meadows - C - New species for Molise region (Mainarde).

A ***Fallopia baldschuanica*** (Regel) Holub (*F. aubertii* (L. Henry) Holub) - uncultivated land - INV - Gole del Sagittario! (Conti and Tinti 2012).

Fallopia convolvulus (L.) Á. Löve (*Polygonum convolvulus* L.) - cultivated land - C

Fallopia dumetorum (L.) Holub (*Polygonum dumetorum* L.) - edges of the woods - PC

Persicaria amphibia (L.) Delarbre (*Polygonum amphibium* L.) - sheets of water - PC

Persicaria hydropiper (L.) Delarbre (*Polygonum hydropiper* L.) - humid environments - RR - near Castel S. Vincenzo (Buchwald 1995).

Persicaria lapathifolia (L.) Delarbre subsp. ***lapathifolia*** - humid environments - C

Polygonum arenastrum Boreau subsp. ***arenastrum*** - trampled uncultivated land - C

Polygonum aviculare L. subsp. ***aviculare*** - abandoned fields, ruderal environments - C

Polygonum aviculare L. subsp. ***rurivagum*** (Jord. ex Boreau) Berher (*P. rurivagum* Jord. ex Boreau) - fields, ruderal environments, pastures - PC

Polygonum bellardii All. (*P. patulum* auct.) - fields - PC

Rumex acetosa L. subsp. ***acetosa*** - pastures - C

Rumex acetosella L. subsp. ***angiocarpus*** (Murb.) Murb. (incl. *R. acetosella* L. subsp. *pyrenaicus* (Pourr. ex Lapeyr.) Akeroyd; *R. angiocarpus* Murb.) - pastures, arid uncultivated land - C

Rumex alpinus L. (*R. pseudoalpinus* Höfft) - humid places where animals gather - C

Rumex arifolius All. (*R. amplexicaulis* Lapeyr.; *R. alpestris* auct. Fl. Ital.) - glades of the *Fagus sylvatica* wood and cool and fertilized meadows - C

Rumex bucephalophorus L. - NC - below S. Biagio (Zodda 1931).

Rumex confertus Willd. - montane meadows - R - territory of Villavallelonga, Villetta Barrea (Grande 1913; Anzalone and Bazzichelli 1960).

Rumex conglomeratus Murray - humid environments - C

Rumex crispus L. - ruderal environments - C

Rumex hydrolapathum Huds. - NC - S. Biagio at Vallevenafrana (Terracciano 1873).

Rumex intermedius DC. - NC - Picinisco ai Treconfini (Terracciano 1878).

P

Rumex nebroides Campd. - stony pastures - CC
Rumex nepalensis Spreng. - uncultivated land - R - territory di Villavallelonga (Grande 1913; Sipari 1926 sub *R. nepalensis* Spr. var. *grandeanus* Chiov. quotes Pirotta; Lusina 1954 sub *R. nepalensis* Spr. var. *grandeanus* Chiov.).
Rumex obtusifolius L. subsp. ***obtusifolius*** - humid environments - C
Rumex obtusifolius L. subsp. ***sylvestris*** (Wallr.) Čelak. - humid meadows - R - Prati d'Angro, Barrea (Anzalone and Bazzichelli 1960).
A ***Rumex patientia*** L. subsp. ***patientia*** - pastures - NAT
Rumex pulcher L. subsp. ***pulcher*** - fields - R - near the Abbazia of S. Vincenzo! (Conti 1995), Valle dell'Inferno!
Rumex sanguineus L. - cool woods, humid meadows - PC
Rumex scutatus L. subsp. ***scutatus*** - screes - CC

Polypodiaceae

Polypodium cambricum L. (*P. australe* Fée; *P. cambricum* L. subsp. *australe* (Fée) Greuter & Burdet ; *P. cambricum* L. subsp. *serrulatum* (Schinz ex Arcang.) Pic. Serm.; *P. vulgare* L. subsp. *serrulatum* Schinz ex Arcang.) - walls, shady cliffs and woods - C
Polypodium interjectum Shivas - shady cliffs and woods - PC
Polypodium vulgare L. - shady cliffs and woods - PC

Portulacaceae

Portulaca granulatostellulata (Poelln.) Ricceri & Arrigoni (*P. oleracea* L. subsp. *granulato-stellulata* (Poelln.) Danin & H. G. Baker) - fields and ruderal environments - R - Gole del Sagittario! (Conti and Tinti 2012), Vallone Lacerno!
Portulaca oleracea L. subsp. *oleracea* - NC - Mainarde (Zodda 1931).

Potamogetonaceae

Groenlandia densa (L.) Fourr. (*Potamogeton densus* L.) - sluggish or stagnant waters - R - Il Pantano (Pedrotti 1983), La Brionna (Tammaro and Visca 1987), Sorgenti del Volturno! (Conti 1992).
Potamogeton berchtoldii Fieber - stagnant waters - R - La Brionna (Tammaro 1984a, b), Le Forme! (Conti 1995).
Potamogeton cfr. *obtusifolius* Mert. & W. D. J. Koch - D - Laghetto near Rif. Campitelli (Buchwald 1995).
Potamogeton crispus L. - running and stagnant waters - R - Montenero Val Cocchiara! (Conti and Minutillo 2001), Lago Pantaniello!
Potamogeton lucens L. - stagnant waters - RR - Lago Pantaniello! (Naviglio 1984).
Potamogeton natans L. - stagnant waters - C

Potamogeton nodosus Poir. - stagnant waters - sources - RR - Capo Volturno (Buchwald 1995).

Potamogeton polygonifolius Pourr. - stagnant waters - RR - Lago Vivo (Anzalone and Bazzichelli 1960).

Potamogeton trichoides Cham. & Schltdl. - stagnant waters - R - lago Pantaniello! (Naviglio 1984), Lago Vivo!

****Zannichellia palustris*** L. - stagnant waters - RR - Bosco Frascaro on the right bank of Rio Chiaro, near the confluence with the river Volturno! The specimens collected are incomplete (without fruits) and thus is impossible to identify the subspecies.

Primulaceae

Androsace maxima L. (Fig. 4) - arid meadows, fields - R - Sperone! (Conti 1998), generically recorded for the Park (Sipari 1926 quotes Pirotta; Lusina 1954).

Androsace villosa L. subsp. ***villosa*** - stony mountain tops - C

Cyclamen hederifolium Aiton subsp. ***hederifolium*** - woods - C

Cyclamen repandum Sm. subsp. ***repandum*** - woods - PC

Lysimachia arvensis (L.) U. Manns & Anderb. subsp. ***arvensis*** (*Anagallis arvensis* L. subsp. *arvensis*) - uncultivated land, fields - C

Lysimachia linum-stellatum L. (*Asterolinon linum-stellatum* (L.) Duby) - arid meadows - C

Lysimachia nemorum L. - woods - RR - Valle di Canneto (Minutillo 1995), Fiume Melfa!

Lysimachia vulgaris L. - humid meadows - R - Villetta Barrea, Val Fondillo! (Anzalone and Bazzichelli 1960), Montenero Val Cocchiara!

Fig. 4 *Androsace maxima* (Photo by F. Bartolucci)

Primula auricula L. (*P. auricula* L. subsp. *balbisii* (Lehm.) Nyman; *P. auricula* L. subsp. *bauhini* (Beck) Lüdi; *P. auricula* L. subsp. *ciliata* (Moretti) Lüdi) - cool cliffs - C

Primula veris L. subsp. *suaveolens* (Bertol.) Gutermann & Ehrend. (*P. columnae* Ten.; *P. suaveolens* Bertol.; *P. veris* L. subsp. *columnae* (Ten.) Lüdi) - NC - alla Difensa, Angro (Grande 1904).

Primula vulgaris Huds. subsp. ***vulgaris*** (*P. acaulis* (L.) Hill) - woods - CC

Soldanella alpina L. subsp. ***alpina*** - snowbed meadows, shady terraces - PC

Pteridaceae

Adiantum capillus-veneris L. - humid and wet cliffs - PC

R

Ranunculaceae

Aconitum lycoctonum L. emend. Koelle (incl. *A. lamarckii* Rchb.; incl. *A. lycoctonum* L. emend. Koelle subsp. *neapolitanum* (Ten.) Nyman; incl. *A. lycoctonum* L. emend. Koelle subsp. *vulparia* (Rchb. ex Spreng.) Nyman; *A. neapolitanum* Ten.; incl. *A. lupicida* Rchb.; incl. *A. ranunculifolium* Rchb.) - *Fagus sylvatica* wood glades - PC

Actaea spicata L. - *Fagus sylvatica* woods - PC

Adonis aestivalis L. subsp. *aestivalis* - NC - Picinisco (Tenore and Gussone 1842).

Adonis annua L. (incl. *A. annua* L. subsp. *castellana* (Pau) C. H. Steinb.; incl. *A. annua* L. subsp. *cupaniana* (Guss.) C. H. Steinb.; incl. *A. castellana* Pau; incl. *A. cupaniana* Guss.) - fields - R - Villavallelonga (Anzalone and Bazzichelli 1960), near the sources of Volturno! (Conti 1995).

Adonis flammea Jacq. subsp. *flammea* - NC - Villavallelonga (Steinberg 1971).

Anemonastrum narcissiflorum (L.) Holub subsp. ***narcissiflorum*** (*Anemone narcissiflora* L. subsp. *narcissiflora*) - high meadows, grassy terraces - PC

Anemone apennina L. subsp. ***apennina*** - woods up to 1,600 m - C

Anemone hortensis L. subsp. ***hortensis*** - arid meadows - R - near Rocchetta Nuova, Monte della Rocchetta! (Conti 1995).

Anemonoides nemorosa (L.) Holub (*Anemone nemorosa* L.) - cool woods - C

Anemonoides ranunculoides (L.) Holub - PC

E ***Aquilegia dumeticola*** Jord. (*A. vulgaris* auct. Fl. Ital.; *A. viscosa* auct. Fl. Ital.) - glades and margins of woods - C

E ***Aquilegia magellensis*** F. Conti & Soldano (*A. ottonis* auct.) (Fig. 1) - wet calcareous cliffs - R - Camosciara! (Terracciano 1890 sub *A. pyrenaica*; Anzalone and Bazzichelli 1960; Bruno and Bazzichelli 1966; Pirone and De

F. Conti, F. Bartolucci, *The Vascular Flora of the National Park of Abruzzo, Lazio and Molise (Central Italy)*, Geobotany Studies, DOI 10.1007/978-3-319-09701-5_21

Fig. 1 *Aquilegia magellensis* (Photo by F. Conti)

Nuntiis 2002), Mainarde! (Conti 1992; Giancola and Stanisci 2006), Marsicano! (Conti 1995), near Guado delle Capre! (Conti and Bartolucci 2011a), Gole del Sagittario! (Conti and Tinti 2012).

Caltha palustris L. (*C. cornuta* Schott, Nyman & Kotschy; incl. *C. laeta* Schott, Nyman & Kotschy; incl. *C. palustris* L. subsp. *laeta* (Schott, Nyman & Kotschy) Hegi; *C. palustris* L. subsp. *cornuta* (Schott, Nyman & Kotschy) Hegi) - marshy environments - R - Il Pantano! (Pedrotti 1983), Campitelli, near Alfedena! (Conti 1994).

Ceratocephala falcata (L.) Cramer (incl. *C. falcata* (L.) Cramer subsp. *incurva* (Steven) Chrtek & Chrtková; incl. *C. incurva* Steven; *Ranunculus falcatus* L.) - NC - S. Angelo, S. Elia, Cacchiamele (Grande 1904).

Clematis flammula L. - hedges, maquis up to 600–700 m - PC

Clematis vitalba L. - hedges, woods up to 1,200–1,300 m - C

Delphinium ajacis L. (*Consolida ajacis* (L.) Schur) - fields and uncultivated land - RR - Camosciara (Anzalone and Bazzichelli 1960).

Delphinium consolida L. (*Consolida regalis* Gray) - fields - PC - The subspecies do not seem clearly distinct as we found in the same population individuals with characters related to subsp. *consolida* as well to subsp. *paniculatum*.

Delphinium fissum Waldst. & Kit. subsp. ***fissum*** - *Fagus sylvatica* wood glades - PC

Delphinium halteratum Sm. subsp. ***halteratum*** - arid meadows - RR - Gole del Sagittario (Conti and Tinti 2012).

Delphinium pubescens DC. (*Consolida pubescens* (DC.) Soó) - fields - RR - Collerotondo di Scanno (Tammaro and Visca 1987).

Eranthis hyemalis (L.) Salisb. (*Helleborus hyemalis* L.) - fields - C

Ficaria verna Huds. subsp. ***verna*** (*Ranunculus ficaria* L. subsp. *bulbilifer* Lambinon; *Ficaria verna* Huds. subsp. *bulbifera* Á. Löve & D. Löve; *Ficaria bulbifera* (Á. Löve & D. Löve) Holub; *Ranunculus ficaria* L. subsp. *ficaria*) - fields, uncultivated land, glades, margins of woods - PC

Ficaria verna L. subsp. ***calthifolia*** (Rchb.) Nyman (*Ranunculus ficaria* L. subsp. *calthifolius* (Rchb.) Arcang.; *Ficaria calthifolia* Rchb.; *Ranunculus calthifolius* (Rchb.) Jord.) - stony meadows - PC

Helleborus foetidus L. subsp. ***foetidus*** - coppices and margins of woods - CC

E ***Helleborus viridis*** L. subsp. ***bocconei*** (Ten.) Peruzzi (*Helleborus bocconei* Ten.) - coppices and margins of woods - R - near Trasacco (Anzalone and Bazzichelli 1960; Zanotti and Cristofolini 1994). The reports of *H. niger* are to be referred to this taxon.

Hepatica nobilis Mill. - woods - CC

Myosurus minimus L. - humid muds - RR - Le Forme (Conti 1995).

Nigella damascena L. - fields, arid uncultivated land - PC

Pulsatilla alpina (L.) Delarbre subsp. ***millefoliata*** (Bertol.) D. M. Moser - stony meadows at high altitudes, cool screes - C

Ranunculus acris L. subsp. ***acris*** - humid meadows - C

E ***Ranunculus apenninus*** (Chiov.) Pignatti (*R. montanus* Willd. var. *apenninus* Chiov.) - montane and subalpine pastures - C

Ranunculus arvensis L. - fields - PC

Ranunculus brevifolius Ten. - pebbly areas and crests - PC

Ranunculus breyninus Crantz (*R. oreophilus* Bieb.) - subalpine rocky pastures - PC

Ranunculus bulbosus L. (incl. *R. bulbifer* Jord.; incl. *R. bulbosus* L. subsp. *aleae* (Willk.) Rouy & Foucaud; incl. *R. bulbosus* L. subsp. *adscendens* (Brot.) Neves) - meadows and uncultivated land - C

Ranunculus flammula L. - RR - Il Pantano (Tammaro 1986).

Ranunculus fluitans Lam. - D - Picinisco nelle acque dei canali presso la cappella di Canneto (Terracciano 1873 sub *R. fluviatilis* Wild.).

Ranunculus garganicus Ten. - NC - Picinisco alla Pietrosa (Terracciano 1890).

Ranunculus gramineus L. - pastures - PC

Ranunculus illyricus L. - arid pastures - PC

Ranunculus lanuginosus L. - woods and glades - CC

E ***Ranunculus magellensis*** Ten. (Fig. 2) - grassy terraces and pebbly areas at the cliff feet and exposed to a prolonged snow - PC

E ***Ranunculus marsicus*** Guss. & Ten. (incl. *R. marsicus* var. *incisior* Dunkel) - humid meadows in the karst plains - PC

Ranunculus millefoliatus Vahl - arid pastures - CC

Ranunculus monspeliacus L. subsp. ***monspeliacus*** - pastures - C

Fig. 2 *Ranunculus magellensis* (Photo by F. Conti)

Ranunculus neapolitanus Ten. (*R. bulbosus* L. subsp. *aleae* auct.; *R. bulbosus* L. subsp. *neapolitanus* (Ten.) H. Lindb.) - meadows and uncultivated land - PC

Ranunculus nemorosus DC. (*R. serpens* Schrank subsp. *nemorosus* (DC.) G. López; *R. tuberosus* Lapeyr.) - humid meadows - PC

Ranunculus parviflorus L. (*R. parvifolius* L.) - dripping calcareous cliffs - RR - S. Michele a Foce near the Eremo! (Conti 1995).

E ***Ranunculus pollinensis*** (N. Terracc.) Chiov. (*R. montanus* Willd. var. *pollinensis* N. Terracc.; *R. sartorianus* auct. Fl. Ital.) - montane and subalpine pastures - C

Ranunculus polyanthemoides Boreau (*R. polyanthemos* L. subsp. *polyanthemoides* (Boreau) Ahlfv.) - humid meadows - RR - Bocche Chiarano (Di Pietro et al. 2005).

Ranunculus repens L. - humid meadows, water courses - CC

Ranunculus sardous Crantz (*R. sardous* Crantz subsp. *subdichotomicus* Gerbault) - humid meadows - PC

Ranunculus sect. ***Batrachium*** DC. (*R. aquatilis* auct. Fl. Ital.) - sluggish waters - R - Valico di Pantano (Vaccari and Wilczek 1940), Le Fontane! (Conti 1995), Sorgenti del Volturno!

*E ***Ranunculus thomasii*** Ten. (*R. polyanthemos* L. subsp. *thomasii* (Ten.) Tutin) (Fig. 3) - humid meadows - R - Bisegna (Anzalone and Bazzichelli 1960 sub

Fig. 3 *Ranunculus thomasii* (Photo by F. Conti)

R. polyanthemos L.), Templo! (Conti 1995 sub *R. polyanthemos* L. s.l.), Padura!, La Ciceranа!

Ranunculus thora L. - cool stony slopes - PC

Ranunculus trichophyllus Chaix subsp. ***trichophyllus*** - stagnant and sluggish waters - PC

Ranunculus velutinus Ten. - humid meadows - R - Il Pantano (Pedrotti 1983), Piana di Pescasseroli (Pedrotti et al. 1992), Sorgenti del Volturno! (Conti 1995).

Thalictrum aquilegiifolium L. subsp. ***aquilegiifolium*** - *Fagus sylvatica* wood glades - CC

Thalictrum flavum L. (*Th. exaltatum* sensu Pignatti; *Th. morisonii* C. C. Gmel.) - D - Valle di Canneto (Tenore and Gussone 1842; Petriglia 2004), generically recorded for the Park (Anzalone and Bazzichelli 1960).

Thalictrum lucidum L. (*Th. mediterraneum* Jord.; *Th. morisonii* C. C. Gmel. subsp. *mediterraneum* (Jord.) P. W. Ball) - riversides and ditches - R - near the lake of Castel S. Vincenzo!, Cartiera!, Sorgenti del Volturno! (Conti 1995), Montenero Val Cocchiara!

Thalictrum minus L. subsp. ***minus*** - stony glades - PC

Thalictrum simplex L. subsp. ***simplex*** - humid meadows especially in the karst plains - PC

Trollius europaeus L. subsp. ***europaeus*** - glades in the upper part of the *Fagus sylvatica* woods, grassy terraces - PC

Resedaceae

Reseda lutea L. subsp. ***lutea*** - uncultivated land and ruins - PC
Reseda luteola L. - stony uncultivated land - PC
Reseda phyteuma L. subsp. ***phyteuma*** - uncultivated land and ruins - PC

Rhamnaceae

Paliurus spina-christi Mill. (*P. australis* Gaertn.; *Rhamnus paliurus* L.) - arid slopes - RR - Gole del Sagittario! (Conti and Tinti 2012).
Rhamnus alaternus L. subsp. ***alaternus*** - *Quercus ilex* groves and Mediterranean maquis - RR - Monte la Rocca near Picinisco (Spada 1979).
Rhamnus alpina L. subsp. ***alpina*** - screes, stony glades within *Fagus sylvatica* woods, to 1,900 m - C
Rhamnus alpina L. subsp. ***fallax*** (Boiss.) Maire & Petitm. (*Rh. fallax* Boiss.) - screes, stony glades within *Fagus sylvatica* woods, to 1,900 m - C
Rhamnus cathartica L. - woods, scrub - PC
Rhamnus pumila Turra - calcareous cliffs - C
Rhamnus saxatilis Jacq. - Mediterranean and supramediterranean arid stony slopes - C - the distinction between the subsp. *saxatilis* and subsp. *infectoria* (L.) P. Fourn. is unclear and worths further investigation.

Rosaceae

Agrimonia eupatoria L. subsp. ***eupatoria*** - arid meadows, uncultivated land - C
Alchemilla alpina L. aggr. (incl. *A. debilicaulis* Buser) - pastures at high altitude - C
Alchemilla cfr. *inconcinna* Buser - D - Passo Godi! La Metuccia!
**Alchemilla* cfr. *cinerea* Buser - pastures - D - Vallone Lampazzo!
Alchemilla colorata Buser (incl. *A. illyrica* Rothm.) - pastures - C
Alchemilla coriacea Buser - pastures - PC
****Alchemilla exigua*** Buser ex Paulin - pastures - RR - Vallone del Carapale - Le Ciminiere!
****Alchemilla filicaulis*** Buser (incl. *A. filicaulis* Buser subsp. *vestita* (Buser) M. E. Bradshaw) - pastures - RR - Monte Cavallo!
Alchemilla flabellata Buser - pastures at high altitude - R - Casalorda (Settefrati) (Conti 1995), Posta Chiarano (Di Pietro et al. 2005), Valle Pagana!
Alchemilla glabra Neygenf. - pastures - RR - Valle di Chiarano! (Conti 1998).
Alchemilla glaucescens Wallr. - pastures - PC

Alchemilla heteropoda Buser - pastures - R - Valle Venafrana! Some specimens collected at Camosciara, Valle Fredda, Le Forme and Valle Ura are doubtfully referred to this taxon by G. Tondi.

E ***Alchemilla marsica*** Buser - pastures - R - Villavallelonga (Vaccari 1911). Some specimens collected at Valle dell'Inferno and M. Marsicano are doubtfully referred to this taxon by G. Tondi.

Alchemilla monticola Opiz - pastures - R - loc. Forestella (Conti 1995; Petriglia 2004), La Metuccia! The reports of *A. vulgaris* var. *sylvestris* for Rifugio della Difesa near Pescasseroli, Villavallelonga and Tre Confini (Anzalone and Bazzichelli 1960) are to be referred to this species.

Alchemilla nitida Buser - calcareous cliffs - PC

Alchemilla sinuata Buser - D - La Metuccia, Monte a Mare (Conti 1995), Picinisco (Petriglia 2004). Probably not present in Italy (Festi in litt.).

Alchemilla straminea Buser - pastures - RR - Val Fondillo! (Conti and Minutillo 1998).

Alchemilla subcrenata Buser - pastures - RR - Valle di Canneto (Minutillo 1995).

Alchemilla transiens (Buser) Buser - D - Monte Greco (Di Pietro et al. 2004), Picinisco (Petriglia 2004).

Alchemilla undulata Buser - pastures - PC

Alchemilla xanthochlora Rothm. - pastures - PC

Amelanchier ovalis Medik. subsp. ***ovalis*** - scrub on cliff, open woods to 1,200 m - PC - Vaccari and Wilczek (1940 sub *A. vulgaris* var. *cretica* DC.) reported *A. ovalis* subsp. *cretica* for the surroundings of Gioia Vecchio but according Anzalone and Bazzichelli (1960) it is doubtfully present.

Aphanes arvensis L. - arid pastures, fields - PC

Aremonia agrimonoides (L.) DC. subsp. ***agrimonoides*** - *Orno - ostryetum* formations, *Fagus sylvatica* woods - CC

Cotoneaster integerrimus auct. Fl. Ital. (*C. mathonnetii* Gand.) - rocky places - PC

Cotoneaster tomentosus (Aiton) Lindl. (*C. nebrodensis* auct., non (Guss.) K. Koch; *Mespilus tomentosa* Aiton) - rocky places - C

Crataegus germanica (L.) Kuntze (*Mespilus germanica* L.) - subacidic broad - leaved woods - RR - Casale! (Conti 1992).

Crataegus laevigata (Poir.) DC. (*C. oxyacantha* L.) - deciduous broad - leaved woods and glades - C

Crataegus monogyna Jacq. (*C. azarella* (Griseb.) Franco) - margins of woods, hedges - C

A ***Cydonia oblonga*** Mill. - roadsides - NAT

Dasiphora fruticosa (L.) Rydb. (*Potentilla fruticosa* L.) - D - generically recorded for the Park (Sipari 1926 quotes Pirotta; Lusina 1954).

Dryas octopetala L. subsp. ***octopetala*** - pebbly slopes on the mountain tops, stony slopes at high altitudes - PC

Filipendula ulmaria (L.) Maxim. (*F. denudata* (J. Presl & C. Presl) Fritsch; incl. *F. ulmaria* (L.) Maxim. subsp. *denudata* (J. Presl & C. Presl) Hayek) - humid meadows - PC

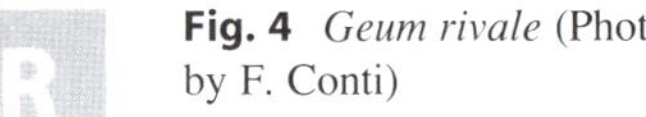

Fig. 4 *Geum rivale* (Photo by F. Conti)

Filipendula vulgaris Moench (*Spiraea filipendula* L.) - NC - Angro, alla Difensa (Grande 1904).

Fragaria vesca L. subsp. ***vesca*** - glades, open woods - CC

Geum molle Vis. & Pančić - montane pastures - C

Geum rivale L. (Fig. 4) - marshy meadows - R - Pantano di Scanno (Avena and Rosati 1974), Villavallelonga (Del Prete et al. 1981 from a specimen collected by Grande), Campitelli! (Conti 1994).

Geum urbanum L. (*G. micropetalum* Gasp.) - glades, coppices - CC

Malus florentina (Zuccagni) C. K. Schneid. (*Crataegus florentina* Zuccagni; *Pyrus florentina* (Zuccagni) O. Targ. Tozz.; *Sorbus florentina* (Zuccagni) K. Koch) - NC - Villavallelonga (Grande 1924).

A ***Malus pumila*** Mill. (*Pyrus malus* L.; *Malus domestica* Borkh., nom. illeg.) - margins and glades of woods - NAT

Malus sylvestris (L.) Mill. - margins and glades of woods - PC

Potentilla apennina Ten. subsp. ***apennina*** (Fig. 3 in chapter "Vegetation Features") - cliffs at high altitude - PC

Potentilla brauneana Hoppe - damp snowbed meadows - R - la Metuccia - Monte a Mare! (Conti 1992), Monte Tartaro!

Potentilla caulescens L. subsp. ***caulescens*** - cliffs - C

Potentilla crantzii (Crantz) Beck ex Fritsch subsp. ***crantzii*** - snowbed meadows, pastures at high altitudes - C

Potentilla detommasii Ten. - pastures - C

Potentilla erecta (L.) Raeusch. - humid meadows - R - Gioia Vecchio - Pescasseroli, Civitella Alfedena (Anzalone and Bazzichelli 1960), Camosciara!

Potentilla incana G. Gaertn., B. Mey. & Scherb. (*P. arenaria* auct. Fl. Ital.) - stony pastures - R - Collelongo, western slopes of the Ciocco (Grande 1913 sub *P. praetutiana* (Ten.) Grande), lower Valle Macrana!

Potentilla inclinata Vill. (*P. canescens* Besser) - pastures - RR - Rifugio della Difesa near Pescasseroli (Anzalone and Bazzichelli 1960 as *P. argentea* L. *canescens* (Besser)).

Potentilla micrantha Ramond ex DC. - woods, glades - C

Potentilla pedata Willd. ex Hornem. (*P. hirta* L. subsp. *laeta* auct. Fl. Ital.; *P. hirta* auct. Fl. Ital.) - arid meadows, pastures - C

Potentilla recta L. subsp. ***recta*** - pastures, uncultivated land - PC

Potentilla reptans L. - humid meadows, banks, ditches - C

E ***Potentilla rigoana*** Th. Wolf - stony pastures and meadows at high altitude - C

Potentilla thuringiaca Bernh. ex Link (*P. chrysantha* Rchb. [non Trevir.]) - NC - Villavallelonga, below the Coppa di Selva Bella (Grande 1913 sub *P. chrysantha* Trev. var. *normalis* Th. Wolf). Generically recorded for the Park (Sipari 1926 quotes Pirotta; Lusina 1954).

Poterium sanguisorba L. subsp. ***balearicum*** (Bourg. ex Nyman) Stace (*Sanguisorba minor* Scop. subsp. *muricata* (Gremli) Briq.; *S. minor* Scop. subsp. *balearica* (Bourg. ex Nyman) Muñoz Garm. & C. Navarro) - arid meadows - C

Poterium sanguisorba L. subsp. ***sanguisorba*** (*Sanguisorba minor* Scop. subsp. *minor*) - arid meadows - R - Monte Rocchetta!, Gole del Sagittario! (Conti and Tinti 2012).

Prunus avium L. - woods and margins of woods - PC

A ***Prunus cerasifera*** Ehrh. - margins of woods - NAT

A ***Prunus cerasus*** L. - margins of woods - CAS

A ***Prunus domestica*** L. subsp. ***domestica*** - hedges - NAT

A ***Prunus dulcis*** (Mill.) D. A. Webb (*Amygdalus dulcis* Mill.) - abandoned fields - NAT

Prunus mahaleb L. subsp. ***mahaleb*** - rocky places, hedges, margins of the thermophilous woods - C

A *Prunus serotina* Ehrh. - CAS

Prunus spinosa L. subsp. ***spinosa*** - bright woods, scrub, hedges, to 1,500 m - CC

Pyracantha coccinea M. Roem. - margins of woods, glades, hedges - PC

A ***Pyrus communis*** L. subsp. ***communis*** - scrub, broad - leaved woods to 1,400 m - NAT

Pyrus communis L. subsp. ***pyraster*** (L.) Ehrh. - scrub, broad - leaved woods to 1,400 m - PC

****Pyrus cordata*** Desv. - broad - leaved woods - R - Vallone Lampazzo!, Valle di Mezzo (Mainarde)! New species for Molise region.

Rosa agrestis Savi - scrub - PC
Rosa arvensis Huds. - open woods and scrub - PC
Rosa balsamica Besser (*R. obtusifolia* Desv.; *R. tomentella* Léman) - ruderal environments - R - Villetta Barrea, Valle di Canneto (Anzalone and Bazzichelli 1960).
Rosa canina L. (incl. *R. andegavensis* Bastard; incl. *R. nitidula* auct. Fl. Ital.; incl. *R. squarrosa* (A. Rau) Boreau) - open woods, scrub, hedges from the plain to 1,500 m - CC
Rosa corymbifera Borkh. (incl. *R. corymbifera* Borkh. var. *deseglisei* (Boreau) Christ; incl. *R. deseglisei* Boreau) - open woods, scrub, hedges - PC
Rosa dumalis Bechst. - scrub - RR - Pescasseroli (Anzalone and Bazzichelli 1960).
A ***Rosa foetida*** Herrm. - scrub - CAS
Rosa glauca Pourr. (*R. rubrifolia* Vill.) - NC - surroundings of Gioia Vecchio (Vaccari and Wilczek 1940).
Rosa micrantha Borrer ex Sm. - scrub, hedges - RR - Valle di Canneto (Anzalone and Bazzichelli 1960).
Rosa montana Chaix - wood glades - PC
Rosa pendulina L. - stony slopes up to 2,000 m - CC
Rosa pouzinii Tratt. - scrub, hedges - R - Valle della Melfa (Terracciano 1873), S. Gennaro!, Morrone delle Rose!
Rosa rubiginosa L. - NC - Picinisco (Tenore and Gussone 1842; Terracciano 1873), Campoli Apennino (Falqui 1899).
Rosa sempervirens L. - maquis, *Quercus ilex* woods, hedges, from the sea level to 800–900 m - PC
Rosa spinosissima L. (*R. pimpinellifolia* L.) - scrub on stony slopes - R - Vallefoglia (Grande 1904), Civita d'Antino!
Rosa subcanina (Christ) Vuk. (*R. reuteri* L. f. *subcanina* Christ) - scrub - RR - over Castrovalva! (Conti and Tinti 2012).
Rubus caesius L. - humid shady environments, especially in riparian woods - PC
Rubus canescens DC. (*R. candicans* auct. Fl. Ital. p.p.) - hedges and open woods - C
Rubus hirtus (group) (incl. *R. hirtus* Waldst. & Kit.; incl. *R. glandulosus* Bellardi) - *Fagus sylvatica* woods - CC
Rubus idaeus L. subsp. ***idaeus*** - glades of montane and subalpine woods, screes - PC
Rubus saxatilis L. - screes generally in the higher part of the *Fagus sylvatica* wood - PC
Rubus ulmifolius Schott (incl. *R. dalmatinus* Tratt. ex Focke; *R. discolor* Weihe & Nees) - hedges, woods, uncultivated land, from the plain up to 1,000–1,200 m - CC
Sanguisorba officinalis L. - NC - surroundings of Gioia Vecchio (Vaccari and Wilczek 1940).
Sibbaldia procumbens L. - damp snowbed meadows, plains at high altitude - PC
Sorbus aria (L.) Crantz subsp. ***aria*** - open woods, stony slopes, glades, mostly in the mountain belt - CC

Sorbus aria (L.) Crantz subsp. ***cretica*** (Lindl.) Holmboe (*S. graeca* (Lodd. ex Spach) Kotschy; *S. meridionalis* (Guss.) Fritsch; *Crataegus graeca* Lodd. ex Spach) - open woods, stony slopes, glades, mostly in the mountain belt - R - Civitella Alfedena, Camosciara (Loche et al. 1992).

Sorbus aucuparia L. (*Pyrus aucuparia* (L.) Gaertn.) - glades in the *Fagus sylvatica* wood and subalpine scrub - C

Sorbus chamaemespilus (L.) Crantz (*Pyrus chamaemespilus* (L.) Ehrh.) - subalpine stony slopes - PC

Sorbus domestica L. - supramediterranean scrub to about 800 m - PC

Sorbus hybrida L. (*Sorbus aucuparia* L. subsp. *hybrida* (L.) Bonnier & Layens) - scrub - RR - Val di Corte (Petriccione 1988).

Sorbus intermedia (Ehrh.) Pers. (*Pyrus intermedia* Ehrh.; *Sorbus mougeotii* Soy.-Will. & Godr.; *Sorbus intermedia* (Ehrh.) Pers.; *Sorbus austriaca* (Beck) Prain) - upper margins of *Fagus sylvatica* woods - R - Vallone della Terratta! (Conti 1998), Camosciara, Scatafosse!

Sorbus torminalis (L.) Crantz (*Pyrus torminalis* (L.) Ehrh.) - open woods - PC

*A ***Spiraea hypericifolia*** L. subsp. ***obovata*** (Waldst. & Kit. ex Willd.) H. Huber - hedges, roadsides, stony slopes - NAT

Rubiaceae

Asperula aristata L. f. subsp. ***aristata*** (*A. aristata* L. f. subsp. *scabra* auct. p. p. majore; *A. scabra* auct. p. p. majore; *A. aristata* L. f. subsp. *longiflora* auct. p. p. majore; *A. longiflora* auct. p. p. majore; *A. flaccida* Ten.; *A. longiflora* Waldst. & Kit. subsp. *flaccida* (Ten.) Nyman) - stony slopes, cliffs - PC

Asperula aristata L. f. subsp. ***oreophila*** (Briq.) Hayek (*A. scabra* C. Presl, nom. illeg.; *A. longiflora* Waldst. & Kit., nom. rej. prop.; *A. aristata* L. f. subsp. *longiflora* (Waldst. & Kit.) Hayek, nom. rej. prop.; *A. aristata* L. f. subsp. *scabra* Nyman, nom. rej. prop.) - stony slopes, cliffs - C

Asperula arvensis L. - fields, uncultivated land - PC

Asperula cynanchica L. subsp. ***cynanchica*** - arid pastures, stony areas - PC

E ***Asperula cynanchica*** L. subsp. ***neglecta*** (Guss.) Arcang. - stony meadows at high altitude - PC

Asperula laevigata L. - *Fagus sylvatica* woods - PC

Asperula purpurea (L.) Ehrend. - stony pastures, stony slopes - C

Asperula taurina L. subsp. ***taurina*** - *Fagus sylvatica* woods - PC

Crucianella angustifolia L. - arid pastures - R - Villavallelonga (Anzalone and Bazzichelli 1960), Cocullo (Greco and Petriccione 1989).

Cruciata glabra (L.) C. Bauhin ex Opiz (*C. glabra* (L.) Ehrend. subsp. *hirticaulis* (Beck) Natali & Jeanm. ; *C. verna* (Scop.) Gutermann & Ehrend.; *Valantia glabra* L.; *Galium vernum* Scop.) - hedges, scrub - C

Cruciata laevipes Opiz - hedges, margins of woods, uncultivated land - C

Cruciata pedemontana (Bellardi) Ehrend. - wood glades, pastures - PC

Galium anisophyllon Vill. - stony slopes - C

Fig. 5 *Galium magellense* (Photo by F. Conti)

Galium aparine L. - ruderal environments - C

Galium corrudifolium Vill. - garigue, stony slopes - C - the reports of *G. lucidum* are probably to be referred to this species.

Galium debile Desv. - marshy environments - RR - Sorgenti del Volturno! (Conti 1995).

Galium divaricatum Lam. - uncultivated land - R - Picinisco (Tenore and Gussone 1842 sub *G. microspermum*), Villavallelonga (Anzalone and Bazzichelli 1960), La Cicerana!

E ***Galium magellense*** Ten. (*G. pusillum* L. subsp. *magellense* (Ten.) Nyman) (Fig. 5) - screes - C

Galium mollugo L. subsp. ***erectum*** Syme (*G. album* Mill.) - hedges, scrub - C

Galium mollugo L. subsp. ***mollugo*** (*G. elatum* Thuill.) - hedges, scrub - C

Galium odoratum (L.) Scop. - *Fagus sylvatica* woods - PC

E ***Galium pallidum*** C. Presl (*G. aetnicum* auct. Fl. Ital. p.p.) (Fig. 6) - stony slopes - R - Camosciara! (Conti et al. 2006), Pietre Rosse!

Galium palustre L. subsp. ***elongatum*** (C. Presl) Lange (*G. elongatum* C. Presl) - marshy environments - RR - Gole del Sagittario (Conti and Tinti 2012).

Galium palustre L. subsp. ***palustre*** - marshy environments - C

Galium parisiense L. - arid pastures, garigue - PC

Galium rotundifolium L. subsp. ***rotundifolium*** - cool *Fagus sylvatica* woods - PC

****Galium spurium*** L. (*G. aparine* L. subsp. *spurium* (L.) Hartm.; *Aparine spuria* (L.) Fourr.) - fields, uncultivated land - R - Settefrati!, Vallone Lampazzo!

Galium tricornutum Dandy - fields - PC

Galium verrucosum Huds. subsp. ***verrucosum*** - fields - R - Mainarde (Conti 1995).

Galium verticillatum Danthoine - stony slopes - R - Villavallelonga (Grande 1910), over Castrovalva!

Galium verum L. subsp. ***verum*** - arid pastures, scrub - CC

Rubia peregrina L. (*R. peregrina* L. subsp. *longifolia* (Poir.) O. Bolòs; *R. peregrina* L. subsp. *requienii* (Duby) Cardona & Sierra) - maquis, hedges, woods, to 1,000 m - PC

A ***Rubia tinctorum*** L. - ruderal environments, hedges - NAT

Fig. 6 *Galium pallidum* (Photo by F. Conti)

Sherardia arvensis L. - ruderal environments, garigue, arid uncultivated land - C

Theligonum cynocrambe L. - uncultivated land, ruderal environments - RR - near Campoli Appennino (Conti 1995).

Rutaceae

Dictamnus albus L. (Fig. 7) - scrub, mostly open *orno - ostryetum* formations - PC

Ruta chalepensis L. - D - generically recorded for the Park (Rovesti and Rovesti 1934).

Ruta graveolens L. (*Ruta divaricata* Ten.; *Ruta graveolens* L. subsp. *divaricata* (Ten.) P. Fourn.) - thermophilous stony slopes - PC

Fig. 7 *Dictamnus albus* (Photo by F. Conti)

S

Salicaceae

Populus alba L. - watercourses and clayey uncultivated land mostly on the plains and the hills - C

Populus canescens (Aiton) Sm. - watercourses - R - San Donato Val di Comino, loc. Le Foci (Spada 1979).

Populus nigra L. subsp. ***nigra*** (*P. nigra* L. var. *italica* Münchh.) - watercourses to 1,500 m - C

Populus nigra L. subsp. ***neapolitana*** (Ten.) Asch. & Graebn. - watercourses to 1,500 m - C

Populus tremula L. - glades and edges of cool woods mostly in the montane belt - PC

A ***Populus x canadensis*** Moench - watercourses - NAT - confluence of the Rio della Valle in Lago di Barrea (Spada 1979 as *P. euroamericana*).

Salix alba L. (incl. *S. alba* L. subsp. *coerulea* (Sm.) Rech. f.; incl. *S. alba* L. var. *coerulea* (Sm.) Sm.; incl. *S. alba* L. subsp. *vitellina* (L.) Arcang.; incl. *S. alba* L. var. *vitellina* (L.) Ser.) - watercourses - CC

Salix amplexicaulis Bory - watercourses and humid environments of the submediterranean and montane belt - R - Castrovalva! Il Lagozzo!, Cresta dell'Altare (Conti 1995).

Salix apennina A. K. Skvortsov - watercourses, marshy environments from the hill to the montane belt - CC

Salix breviserrata Flod. - humid grassy slopes of alpine and subalpine belt - RR - Monte Marsicano (Conti 1995).

Salix caprea L. - borders of woods and glades to 1,800 m - C

Salix cinerea L. - watercourses, marshy environments from the plains to 1,300 m - PC

F. Conti, F. Bartolucci, *The Vascular Flora of the National Park of Abruzzo, Lazio and Molise (Central Italy)*, Geobotany Studies, DOI 10.1007/978-3-319-09701-5_22

Salix eleagnos Scop. subsp. **eleagnos** - watercourses of the hill and montane belt - PC

Salix pentandra L. - marshy environments - RR - Montenero Val Cocchiara! (Conti et al. 1992).

Salix purpurea L. subsp. **purpurea** - mostly spread in the watercourses and in the humid environments of the montane areas - C

Salix retusa L. - cool and cliff environments at high altitude - PC

Salix triandra L. subsp. **triandra** (*S. triandra* L. subsp. *amygdalina* (L.) Schübl. & G.Martens ; *S. triandra* L. subsp. *discolor* (Wimm. & Grab.) Arcang.) - humid places, watercourses - RR - Sorgenti del Volturno! (Conti 1995).

Santalaceae

Osyris alba L. - maquis, garigue and thermophilous woods to 800–1,000 m - PC

Thesium alpinum L. (incl. *Th. alpinum* L. subsp. *tenuifolium* (Saut.) O. Schwartz; incl. *T. tenuifolium* Saut.) - stony pastures - PC

Thesium humifusum DC. (*Th. divaricatum* Jan ex Mert. & W. D. J. Koch) - arid meadows - PC

Thesium linophyllon L. - stony pastures and cliffs - C

Thesium parnassi A. DC. - subalpine pastures - C

Viscum album L. subsp. **album** - hemiparasite on different tree species to 1,200 m - PC

Sapindaceae

Acer × peronai Schwer. - woods - R - Collelongo, Villavallelonga (Anzalone and Bazzichelli 1960).

Acer campestre L. (*A. campestre* L. subsp. *marsicum* (Guss.) Hayek; *A. marsicum* Guss.) - thermophilous woods, margins of woods, hedges - CC

E **Acer cappadocicum** Gled. subsp. **lobelii** (Ten.) A. E. Murray (*A. lobelii* Ten.; *A. platanoides* L. var. *lobelii* (Ten.) Loudon) - mesophilous woods between *Quercus cerris* and *Fagus sylvatica* woods - PC

Acer monspessulanum L. subsp. **monspessulanum** - *Quercus ilex* woods and supramediterranean woods - PC

Acer opalus Mill. subsp. **obtusatum** (Waldst. & Kit. ex Willd.) Gams (*A. neapolitanum* Ten.; *A. obtusatum* Waldst. & Kit. ex Willd.; *A. obtusatum* Waldst. & Kit. ex Willd. subsp. *neapolitanum* (Ten.) Pax) - woods with formations of evergreen sclerophylls to the *Fagus sylvatica* woods - C

Acer platanoides L. - cool montane woods to 1,500 m - PC

Acer pseudoplatanus L. - submontane and montane woods to the upper limit of the *Fagus sylvatica* woods at 1,800–1,900 m - PC

A **Aesculus hippocastanum** L. - woods - CAS

Saxifragaceae

Chrysosplenium alternifolium L. - NC - Monte Meta at Forestella of Picinisco (Grande 1916 from a specimen collected by Tenore and Gussone).

Saxifraga adscendens L. subsp. ***adscendens*** - stony slopes - PC

Saxifraga adscendens L. subsp. ***parnassica*** (Boiss. & Heldr.) Hayek - stony slopes - PC

Saxifraga bulbifera L. - pastures and glades - C

Saxifraga caesia L. - cliffs at high altitude - PC

Saxifraga callosa Sm. subsp. ***callosa*** (incl. *S. australis* Moric.; *S. lingulata* Bellardi) - cliffs - C

E ***Saxifraga exarata*** Vill. subsp. ***ampullacea*** (Ten.) D. A. Webb (*S. ampullacea* Ten.) (Fig. 1) - cliffs at high altitude - C

Saxifraga glabella Bertol. - stony slopes subjected to long periods of snow - R - Forca Resuni!, Mainarde! (Anzalone and Bazzichelli 1960; Conti 1992).

Saxifraga granulata L. subsp. ***granulata*** - arid and stony pastures - C

Saxifraga marginata Sternb. - D - Aremogna - Toppe del Tesoro (Pirone 1997). The real presence should be checked because it is not supported by herbarium specimens.

Saxifraga oppositifolia L. subsp. ***oppositifolia*** (incl. *S. latina* (N. Terracc.) Hayek) - stony areas, screes at high altitude - PC

E ***Saxifraga oppositifolia*** L. subsp. ***speciosa*** (Dörfl. & Hayek) Engl. and Irmsh. (*S. speciosa* Dörfl. & Hayek) - stony areas, screes at high altitude - PC

Saxifraga paniculata Mill. (*S. paniculata* Mill. subsp. *stabiana* (Ten.) Pignatti) - cliffs and stony slopes - C

E ***Saxifraga porophylla*** Bertol. subsp. ***porophylla*** (Fig. 2) - cliffs from 1,000 m upwards - C

Saxifraga rotundifolia L. subsp. ***rotundifolia*** - woods - C

Saxifraga tridactylites L. - cliffs and walls - C

Fig. 1 *Saxifraga exarata* subsp. *ampullacea* (Photo by F. Conti)

Fig. 2 *Saxifraga porophylla* subsp. *porophylla* (Photo by F. Conti)

Scrophulariaceae

Scrophularia canina L. subsp. ***bicolor*** (Sm.) Greuter (*S. bicolor* Sm.) - thermophilous woods, scrub, stony areas - PC

Scrophularia hoppii W. D. J. Koch (*S. canina* L. subsp. *hoppii* (W. D. J. Koch) P. Fourn.; *S. juratensis* Schleicher) - screes - PC

Scrophularia nodosa L. - woods - PC

Scrophularia peregrina L. - ruderal environments - R - Picinisco (Tenore and Gussone 1842; Petriglia 2004), Mainarde (Zodda 1931).

Scrophularia scopolii Hoppe ex Pers. - glades - C

Scrophularia umbrosa Dumort. subsp. ***umbrosa*** (*S. aquatica* L.) - ditches, riversides - PC - the reports of *S. auriculata* for Villetta Barrea, Scanno (Anzalone and Bazzichelli 1960 sub *S. aquatica* var. *balbisii*) and Piana di Pescasseroli (Pedrotti et al. 1992) are probably to be referred to this taxon.

Scrophularia vernalis L. - shady cliffs in the *Fagus sylvatica* woods - PC

Verbascum alpinum Turra (*V. lanatum* auct.) - uncultivated land - R - Collelongo, Valle di Canneto (Anzalone and Bazzichelli 1960).

E *Verbascum argenteum* Ten. - NC - Villavallelonga (Anzalone and Bazzichelli 1960 da Grande).

Verbascum blattaria L. - NC - Picinisco (Tenore and Gussone 1842).

Verbascum chaixii Vill. subsp. ***chaixii*** - uncultivated land, glades - R - Collelongo, Villavallelonga (Murbeck 1933; Anzalone and Bazzichelli 1960), Valle Ura!

Verbascum densiflorum Bertol. - pastures - PC

Verbascum longifolium Ten. - pastures - C

Verbascum lychnitis L. - arid pastures, uncultivated areas - PC

Verbascum macrurum Ten. - arid meadows, stony uncultivated land - R - S. Donato Val Comino (Terracciano 1878), Gole del Sagittario! (Conti and Tinti 2012).

Verbascum mallophorum Boiss. & Heldr. - uncultivated land and arid meadows - C

E ***Verbascum niveum*** Ten. subsp. ***garganicum*** (Ten.) Murb. - arid pastures - RR - Villavallelonga (Murbeck 1933; Anzalone and Bazzichelli 1960).

Verbascum phlomoides L. - pastures - PC

Verbascum pulverulentum Vill. - arid pastures - C

Verbascum samniticum Ten. - arid pastures - R - Collelongo (Anzalone and Bazzichelli 1960).

Verbascum sinuatum L. - uncultivated land and arid meadows - PC

Verbascum thapsus L. (*V. crassifolium* DC.; *V. thapsus* L. subsp. *crassifolium* (Lam. & DC.) Murb.; *V. thapsus* L. subsp. *montanum* (Schrad.) Bonnier & Layens) - arid and stony pastures - PC

Selaginellaceae

Selaginella denticulata (L.) Spring (*Lycopodium denticulatum* L.) - humid cliffs - NC - Picinisco (Tenore and Gussone 1842; Terracciano 1873).

Simaroubaceae

A ***Ailanthus altissima*** (Mill.) Swingle (*A. glandulosa* Desf.) - ruderal environments, roadsides, termophilous stony slopes - INV

Smilacaceae

Smilax aspera L. - maquis and thermophilous woods - R - presso Picinisco (Tenore and Gussone 1842; Spada 1979).

Solanaceae

Atropa bella-donna L. - *Fagus sylvatica* wood glades - PC

A ***Datura stramonium*** L. subsp. ***stramonium*** - ruderal environments - NAT

S

Hyoscyamus niger L. - walls - PC

Physalis alkekengi L. - margins of woods - R - Valle di Canneto (Tenore and Gussone 1842), S. Biagio Saracinisco (Terracciano 1872), Valle di Mezzo! (Conti 1995).

Solanum dulcamara L. - uncultivated land, screes, riparian open woods and banks - C

A ***Solanum lycopersicum*** L. (*Lycopersicon esculentum* Mill.) - uncultivated land, riversides - CAS

Solanum nigrum L. (incl. *S. nigrum* subsp. *schultesii* (Opiz) Wessely) - ruderal environments - PC

Solanum villosum Mill. (*S. luteum* Mill.; incl. *S. luteum* Mill. subsp. *alatum* (Moench) Dostal; incl. *S. villosum* Mill. subsp. *alatum* (Moench) Edmonds) - ruderal environments - R - Colleruta! (Conti and Minutillo 1998), Gole del Sagittario (Conti and Tinti 2012), below Castelnuovo!

Staphyleaceae

Staphylea pinnata L. - riversides - R - Fiume Mollarino below S. Biagio! (Conti 1995), impluvium between Collalto Scapoli and Cardito!

T

Taxaceae

Taxus baccata L. - cool *Fagus sylvatica* woods in rocky environments to 1,800 m - PC

Thymelaeaceae

Daphne alpina L. subsp. ***alpina*** - cliffs, pastures, subalpine scrub - PC
Daphne laureola L. - woods to 1,800 m - C
Daphne mezereum L. - woods and margins, scrub - C
Daphne oleoides Schreb. subsp. ***oleoides*** - cliffs, pastures, subalpine scrub - C
Daphne sericea Vahl (Fig. 1) - cliffs and maquis - RR - Gole del Sagittario! (Anzalone 1961; Conti and Tinti 2012).

Typhaceae

Sparganium neglectum Beeby (*S. erectum* L. subsp. *neglectum* (Beeby) Schinz & Thell.) - watercourses, stagnant waters - PC
Typha angustifolia L. - ditches, stagnant waters - PC
Typha domingensis (Pers.) Steud. - ditches, stagnant waters - RR - Torrente Iemmare, presso Pizzone (Conti and Minutillo 2001).
Typha latifolia L. - ditches, stagnant waters - PC
Typha minima Funk ex Hoppe - ditches, stagnant waters - RR - Torrente Iemmare, presso Pizzone (Conti and Minutillo 2001).

F. Conti, F. Bartolucci, *The Vascular Flora of the National Park of Abruzzo, Lazio and Molise (Central Italy)*, Geobotany Studies, DOI 10.1007/978-3-319-09701-5_23

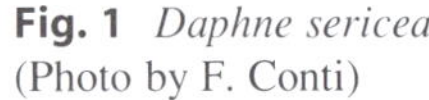
Fig. 1 *Daphne sericea* (Photo by F. Conti)

U

Ulmaceae

Ulmus glabra Huds. - montane mesophile woods - PC

Ulmus minor Mill. subsp. ***minor*** - plain - growing and thermophilous woods of *Quercus pubescens*, *Q. ilex*, *Q. cerris* and *Orno - ostryetum* formations, hedges, uncultivated land - CC

Urticaceae

Parietaria judaica L. (*P. diffusa* Mert. & W. D. J. Koch) - cliffs and sunny ruderal environments - CC

Parietaria officinalis L. - nitrophilous shady environments - PC

Urtica dioica L. subsp. ***dioica*** - ruderal environments and places where animals gather - CC

Urtica urens L. - ruderal environments and places where animals gather - PC

F. Conti, F. Bartolucci, *The Vascular Flora of the National Park of Abruzzo, Lazio and Molise (Central Italy)*, Geobotany Studies, DOI 10.1007/978-3-319-09701-5_24

V

Verbenaceae

Verbena officinalis L. - ruderal environments - CC

Violaceae

Viola alba Besser subsp. ***dehnhardtii*** (Ten.) W. Becker (*V. dehnhardtii* Ten.) - scrub and open woods - CC

Viola arvensis Murray subsp. ***arvensis*** - fields, uncultivated land - PC

E ***Viola eugeniae*** Parl. subsp. ***eugeniae*** - stony pastures - C

E ***Viola eugeniae*** Parl. subsp. ***levieri*** (Parl. ex Caruel) Arcang. (*V. levieri* Parl. ex Caruel) - stony pastures - PC

Viola hirta L. - hedges, meadows - PC

Viola hymettia Boiss. & Heldr. - arid meadows - RR - Gole del Sagittario! (Conti and Tinti 2012).

Viola kitaibeliana Schult. - arid meadows - R - Valle del Giovenco near Rivoli! (Conti et al. 2008), near Scanno!

Viola odorata L. - margins of woods - C

Viola pyrenaica Ramond ex DC. - shady places - RR - near Scanno (Pignatti 1982).

Viola reichenbachiana Jord. ex Boreau - cool woods - CC

Viola riviniana Rchb. - woods - PC

Viola suavis M. Bieb. subsp. ***suavis*** - thermophilous woods - R - Picinisco, Settefrati (Petriglia 2004).

Viola tricolor L. subsp. ***tricolor*** - humid meadows - R - Rif. Aremogna (Hennecke and Hennecke 1999), Valle di Mezzo!

F. Conti, F. Bartolucci, *The Vascular Flora of the National Park of Abruzzo, Lazio and Molise (Central Italy)*, Geobotany Studies, DOI 10.1007/978-3-319-09701-5_25

Vitaceae

Vitis x ***koberi*** Ardenghi, Galasso, Banfi & Lastrucci (*V. berlandieri* Planch. x *V. riparia* Michx.) - INV- The nomenclature according to Ardenghi et al. (2014).

Vitis vinifera L. (*V. vinifera* L. subsp. *silvestris* Hegi; *V. sylvestris* C. C. Gmel., *nom. illeg.*) - watercourses and cool hedges - R

W

Woodsiaceae

Athyrium filix-femina (L.) Roth (*Polypodium filix-femina* L.) - shady woods - C
Cystopteris alpina (Lam.) Desv. (*C. fragilis* (L.) Bernh. subsp. *alpina* (Lam.) Hartman; *Polypodium alpinum* Lam.) - humid cliffs - PC
Cystopteris fragilis (L.) Bernh. (*Polypodium fragile* L.) - humid cliffs - CC
Gymnocarpium dryopteris (L.) Newman (*Polypodium dryopteris* L.) - cool *Fagus sylvatica* woods - R - Camosciara!, Mainarde! (Conti 1995).
Gymnocarpium robertianum (Hoffm.) Newman (*Polypodium robertianum* Hoffm.) - cool *Fagus sylvatica* woods - R - Camosciara!, Villavallelonga! (Grande 1904; Anzalone and Bazzichelli 1960).

F. Conti, F. Bartolucci, *The Vascular Flora of the National Park of Abruzzo, Lazio and Molise (Central Italy)*, Geobotany Studies, DOI 10.1007/978-3-319-09701-5_26

X

Xanthorrhoeaceae

Asphodeline lutea (L.) Rchb. (*A. luteus* L.) - arid and stony slopes - PC

Asphodelus macrocarpus Parl. subsp. ***macrocarpus*** (*A. albus* auct. Fl. Ital.) - montane pastures - C

A ***Hemerocallis fulva*** L. - margins of woods - NAT - Picinisco - Lago di Grotta Campanaro! (Conti 1995).

F. Conti, F. Bartolucci, *The Vascular Flora of the National Park of Abruzzo, Lazio and Molise (Central Italy)*, Geobotany Studies, DOI 10.1007/978-3-319-09701-5_27

Discussion and Conclusion

The above floristic list comprises 2,114 *taxa* (species and subspecies) belonging to 109 families and 680 genera. There is a significant increase of 202 *taxa* compared to the last list of flora (Conti 1995) that reported 1,912 *taxa*. The native flora consists of 2,024 *taxa* which represents a significant number, among the highest recorded in Europe's national parks.

Considering native and non-native *taxa*, the most represented families and genera are reported in Table 1.

Forty nine *taxa* are reported for the first time for the Park flora, 9 *taxa* are new for Molise (*Bidens cernuus*, *Bistorta vivipara*, *Luzula taurica*, *Orobanche rapum-genistae*, *O. teucrii*, *Callitriche lenisulca*, *C. platycarpa*, *Cirsium vulgare* subsp. *crinitum*, and *Pyrus cordata*), 2 are new for Abruzzo (*Bromopsis pannonica* subsp. *pannonica Tilia platyphyllos* subsp. *pseudorubra*) and 1 is confirmed for Molise (*Sedum montanum*).

Species of Phytogeographical and Conservation Interest

The Park is located in the heart of the central Apennines and for this reason is home to a large contingent of plants of floristic interest, with 137 *taxa* endemic to Italy (6.5 %), including 29 endemic to central Italy (Table 2). As also observed for the Italian flora (Peruzzi et al. 2014), the number of endemic species has increased during the last decades and is likely to increase further in the near future.

There are 202 very rare (RR) *taxa*, known only in one locality in the protected area. Some of these are of great interest as they are extremely rare in Italy, such as *Allium permixtum*, known only in the Abruzzo region (Conti 1995) and extinct in Sicily (Brullo et al. 2010), *Gagea ramulosa*, in Italian peninsula, known only in two localities in Abruzzo (Peruzzi and Bartolucci 2006; Bartolucci and Peruzzi 2007), *Ranunculus polyanthemoides* and *Pyrus cordata* in Italy known only in the Abruzzo region, *Pedicularis friderici - augusti* in Italy recorded with certainty only in Abruzzo, *Bufonia paniculata* in Italy recorded only in Sardinia and Abruzzo

F. Conti, F. Bartolucci, *The Vascular Flora of the National Park of Abruzzo, Lazio and Molise (Central Italy)*, Geobotany Studies, DOI 10.1007/978-3-319-09701-5_28

Table 1 Most represented families (>25 *taxa*) and genera (>15 *taxa*)

Families	*taxa*	Genera	*taxa*
Asteraceae	264	*Carex*	46
Poaceae	166	*Hieracium*	35
Fabaceae	160	*Trifolium*	33
Rosaceae	102	*Ranunculus*	28
Brassicaceae	101	*Silene*	23
Apiaceae	95	*Allium*	22
Caryophyllaceae	94	*Veronica*	22
Lamiaceae	92	*Vicia*	22
Orchidaceae	76	*Alchemilla*	20
Ranunculaceae	64	*Galium*	19
Cyperaceae	58	*Ophrys*	19
Plantaginaceae	57	*Rumex*	18
Orobanchaceae	42	*Geranium*	17
Boraginaceae	39	*Cerastium*	16
Rubiaceae	35	*Lathyrus*	16
Caprifoliaceae	34	*Rosa*	16
Polygonaceae	29	*Orobanche*	15
Amaryllidaceae	27	*Saxifraga*	15
Asparagaceae	25	*Verbascum*	15

where it is known to occur in and near the Park, *Saxifraga glabella* and *Scabiosa silenifolia* known with certainty only in Abruzzo and Molise and *Sorbus hybrida* known only in Lombardy and Abruzzo.

The Park represents the southern or northern border of the Italian distribution of 72 *taxa* (Table 3).

The high number of plants at the edge of their Italian distribution area is one of the most interesting results from a phytogeographical point of view and can be explained by a certain continuity with the other mountains of the central Apennines, of which the Park is the southernmost spur. Further south on the other hand, the low and relatively broad valley of the Volturno river which separates the Park from the Matese mountains represents an ecological barrier for many microtherm plants.

Of the *taxa*, 623 are of conservation interest, as endemic, very rare or included in international, national or regional lists. For each *taxa*, Table 4 gives the status in the Red List of Italian Flora (Conti et al. 1997; Rossi et al. 2013) or inclusion in the following regional laws or international conventions: regional laws on the protection of flora no. 61 of 19 September 1974 (Lazio), no. 45 of 11 November 1979 and no. 66 of 20 June 1980 (Abruzzo), no. 9 of 23 February 1999 (Molise), Washington Convention on International Trade in Endangered Species, Habitat Directive 92/43/EEC and Berne Convention on the Conservation of European Wildlife and Natural Habitats. In addition, for the purposes of managing the protected area, we have subdivided the vascular plants thus identified into six protection classes (A, B, C, D, E and F) according to the criteria explained in Table 5.

Table 2 Endemic *taxa* of central Italy

Taxa	Distribution
Alchemilla marsica	Endemic of central Italy
Aquilegia magellensis	Endemic of central Italy
Asperula cynanchica subsp. *neglecta*	Endemic of central Italy
Astrantia pauciflora subsp. *tenorei*	Endemic of central Italy
Biscutella laevigata subsp. *australis*	Endemic of central Italy
Campanula tanfanii	Endemic of central Italy
Centaurea ambigua subsp. *nigra*	Endemic of central Italy
Centaurea ceratophylla subsp. *ceratophylla*	Endemic of central Italy
Cerastium thomasii	Endemic of central Italy
Coristospermum cuneifolium	Endemic of central Italy
Cymbalaria pallida	Endemic of central Italy
Euphorbia gasparrinii subsp. *samnitica*	Endemic of central Italy
Iris marsica	Endemic of central Italy
Leucanthemum tridactylites	Endemic of central Italy
Noccaea stylosa	Endemic of central Italy
Ophrys ausonia	Endemic of central Italy
Orobanche ebuli	Endemic of central Italy
Paeonia officinalis subsp. *italica*	Endemic of central Italy
Pinus nigra subsp. *nigra* var. *italica*	Endemic of central Italy
Ranunculus magellensis	Endemic of central Italy
Ranunculus marsicus	Endemic of central Italy
Saxifraga exarata subsp. *ampullacea*	Endemic of central Italy
Saxifraga oppositifolia subsp. *speciosa*	Endemic of central Italy
Sempervivum riccii	Endemic of central Italy
Senecio ovatus subsp. *stabianus*	Endemic of central Italy
Silene cattariniana	Endemic of central Italy
Stachys alopecuros subsp. *divulsa*	Endemic of central Italy
Taraxacum glaciale	Endemic of central Italy
Viola eugeniae subsp. *eugeniae*	Endemic of central Italy
Viola eugeniae subsp. *levieri*	Endemic of central Italy
Centaurea scannensis	Endemic of Abruzzo
Hieracium latilepidotum	Endemic of Abruzzo
Hieracium marsorum	Endemic of Abruzzo
Hieracium profetanum	Endemic of Abruzzo
Minuartia glomerata subsp. *trichocalycina*	Endemic of Abruzzo
Pilosella corvigena	Endemic of Abruzzo
Pinguicula vallis - regiae	Endemic of the Park

Table 3 Plants at the edge of the Italian distribution area

Taxa	Edge of the distribution
Adonis aestivalis subsp. *aestivalis*	Southern border
Ajuga pyramidalis	Southern border
Alchemilla undulata	Southern border
Allium rotundum	Southern border
Allium schoenoprasum	Southern border
Androsace maxima	Southern border
Arabis bellidifolia subsp. *stellulata*	Southern border
Aster alpinus subsp. *alpinus*	Southern border
Astragalus danicus	Southern border
Bistorta vivipara	Southern border
Carex brachystachys	Southern border
Carex mucronata	Southern border
Carex ornithopoda	Southern border
Carum carvi	Southern border
Cerastium cerastoides	Southern border
Chrysosplenium alternifolium	Southern border
Cirsium oleraceum	Southern border
Colchicum bulbocodium subsp. *versicolor*	Southern border
Cotinus coggygria	Southern border
Cynoglossum officinale	Southern border
Cypripedium calceolus	Southern border
Crepis pygmaea	Southern border
Dactylorhiza maculata subsp. *fuchsii*	Southern border
Dryopteris villarii subsp. *villarii*	Southern border
Dryopteris submontana	Southern border
Epilobium alpestre	Southern border
Epilobium alsinifolium	Southern border
Epipactis helleborine subsp. *orbicularis*	Southern border
Erigeron atticus	Southern border
Euphrasia illyrica	Southern border
Fibigia clypeata	Southern border
Genista radiata	Southern border
Gentiana brachyphylla subsp. *favratii*	Southern border
Gentiana nivalis	Southern border
Gymnocarpium dryopteris	Southern border
Hornungia alpina subsp. *alpina*	Southern border
Hypericum hyssopifolium	Southern border
Iberis saxatilis subsp. *saxatilis*	Southern border
Isatis apennina	Southern border
Klasea nudicaulis	Southern border
Leucanthemum heterophyllum	Southern border
Ophrys riojana	Southern border
Pedicularis hoermanniana	Southern border

(continued)

Table 3 (continued)

Taxa	Edge of the distribution
Pedicularis rostratospicata	Southern border
Pinus mugo subsp. *mugo*	Southern border
Polygala chamaebuxus	Southern border
Polygonatum verticillatum	Southern border
Potentilla brauneana	Southern border
Primula veris subsp. *suaveolens*	Southern border
Rosa glauca	Southern border
Salix breviserrata	Southern border
Saponaria bellidifolia	Southern border
Saxifraga caesia	Southern border
Saxifraga oppositifolia subsp. *oppositifolia*	Southern border
Sibbaldia procumbens	Southern border
Silene acaulis subsp. *bryoides*	Southern border
Silene ciliata subsp. *graefferi*	Southern border
Sorbus chamaemespilus	Southern border
Sorbus intermedia	Southern border
Tephroseris integrifolia subsp. *integrifolia*	Southern border
Traunsteinera globosa	Southern border
Triglochin palustris	Southern border
Trollius europaeus subsp. *europaeus*	Southern border
Vaccinium myrtillus	Southern border
Valeriana saliunca	Southern border
Viola pyrenaica	Southern border
Acer cappadocicum subsp. *lobelii*	Northern border
Euphorbia corallioides	Northern border
Geranium austroapenninum	Northern border
Ophrys lacaitae	Northern border
Seseli peucedanoides	Northern border
Phlomis fruticosa	Northern border

Doubtful and not Confirmed Species

The *taxa* listed include 54 doubtful (2.5 %) species, while 104 (4.9 %) of those previously recorded were not found during the present study. It is worth noting that these *taxa* have decreased compared to those published in the last flora (Conti 1995) which amounted to 188.

Among the doubtful species, some are very likely erroneous as wrongly identified and are therefore likely to be excluded from further study, i.e. *Bunium petraeum*, *Cerastium arvense* subsp. *arvense*, *Rhaponticoides centaurium*, *Alyssum cuneifolium* subsp. *cuneifolium*, *Astragalus muelleri*, *Stachys pradica*,

Table 4 List of *taxa* of conservation interest

Taxa	Classes of protection	Red List of Italian Flora (Rossi et al. 2013)	Regional Red List: Latium (Conti et al. 1997)	Regional Red List: Abruzzo (Conti et al. 1997)	Regional Red List: Molise (Conti et al. 1997)	Regional Law (Latium)	Regional Laws (Abruzzo)	Regional Law (Molise)	International Conventions
Pinguicula vallis-regiae	A (a1, a2, a4)	EN	VU						
Centaurea scannensis	A (a2, a3, a4, a7)			LR			R.L. n°45 - 11/09/1979 and n°66 - 20/06/1980		
Iris marsica	A (a3)	NT	VU	LR					Habitat Directive 92/43/CEE and Berne Convention
Minuartia glomerata subsp. trichocalycina	A (a3, a4)			LR					
Aquilegia magellensis	A (a3, a7)			LR	LR		R.L. n°45 - 11/09/1979 and n°66 - 20/06/1980	R.L. n°9 - 23/02/1999	

Allium permixtum	A (a4, a7)			LR					
Cypripedium calceolus	A (a5, a7)	LC		VU					Habitat Directive 92/43/CEE, Washington Convention and Berne Convention
Traunsteinera globosa	A (a5, a7)			LR					Washington Convention
Salix pentandra	A (a6, a7)		CR		CR			R.L. n°9 - 23/02/ 1999	
Astragalus aquilanus.	A (a7)	EN		VU			R.L. n°45 - 11/09/ 1979 and n°66 - 20/06/ 1980		Habitat Directive 92/43/CEE and Berne Convention
Bufonia paniculata	A (a7)			VU					
Bupleurum rollii	A (a7)					R.L. n°61 - 19/09/ 1987			
Verbascum argenteum	A (a7)			LR	LR			R.L. n°9 - 23/02/ 1999	
Ophrys riojana	B (b2)								

(continued)

Table 4 (continued)

Taxa	Classes of protection	Red List of Italian Flora (Rossi et al. 2013)	Regional Red List: Latium (Conti et al. 1997)	Regional Red List: Abruzzo (Conti et al. 1997)	Regional Red List: Molise (Conti et al. 1997)	Regional Law (Latium)	Regional Laws (Abruzzo)	Regional Law (Molise)	International Conventions
Acer cappadocicum subsp. lobelii	B (b3)		VU	LR	LR			R.L. n°9 - 23/02/ 1999	
Adonis aestivalis subsp. aestivalis	B (b3)			DD	LR			R.L. n°9 - 23/02/ 1999	
Ajuga pyramidalis	B (b3)			LR					
Alchemilla marsica	B (b3)								
Allium calabrum	B (b3)								
Allium rotundum	B (b3)			LR					
Allium schoenoprasum.	B (b3)		LR	LR					
Androsace maxima.	B (b3)			VU					
Arabis bellidifolia subsp. stellulata	B (b3)		LR						
Astragalus danicus	B (b3)			LR					
Carex brachystachys	B (b3)			LR					
Carex mucronata	B (b3)			LR					
Carum carvi	B (b3)		LR	LR					
Cerastium cerastoides	B (b3)		LR	LR	LR			R.L. n°9 - 23/02/ 1999	
Cirsium oleraceum	B (b3)		LR	LR					

Colchicum bulbocodium subsp. versicolor	B (b3)			LR					
Cotinus coggygria	B (b3)			LR					
Crepis pygmaea	B (b3)		LR	LR					
Cymbalaria glutinosa subsp. glutinosa	B (b3)			LR	LR	R.L. n°61 - 19/09/ 1991		R.L. n°9 - 23/02/ 1999	
Cynoglossum officinale.	B (b3)				DD				
Dryopteris submontana	B (b3)			LR					
Epilobium alpestre	B (b3)								
Epilobium alsinifolium	B (b3)		LR						
Epipactis helleborine subsp. orbicularis	B (b3)								Washington Convention
Erigeron atticus	B (b3)			LR	EN			R.L. n°9 - 23/02/ 1999	
Euphorbia gasparrinii subsp. samnitica	B (b3)			LR	EN			R.L. n°9 - 23/02/ 1999	
Fumaria petteri subsp. petteri	B (b3)		LR		LR			R.L. n°9 - 23/02/ 1999	
Gagea ramulosa	B (b3)								
Gentiana nivalis	B (b3)		VU						

(continued)

Table 4 (continued)

Taxa	Classes of protection	Red List of Italian Flora (Rossi et al. 2013)	Regional Red List: Latium (Conti et al. 1997)	Regional Red List: Abruzzo (Conti et al. 1997)	Regional Red List: Molise (Conti et al. 1997)	Regional Law (Latium)	Regional Laws (Abruzzo)	Regional Law (Molise)	International Conventions
Geranium austroapenninum	B (b3)		LR	LR	LR			R.L. n°9 - 23/02/ 1999	
Geum rivale	B (b3)		VU	EN	DD				
Gymnocarpium dryopteris	B (b3)		LR	LR	LR			R.L. n°9 - 23/02/ 1999	
Hornungia alpina subsp. alpina	B (b3)		LR						
Hypericum hyssopifolium	B (b3)		LR	VU	LR			R.L. n°9 - 23/02/ 1999	
Isatis apennina	B (b3)		LR	LR					
Juniperus sabina	B (b3)		LR	LR					
Klasea nudicaulis	B (b3)		VU						
Leucanthemum heterophyllum	B (b3)		LR						
Orobanche ebuli	B (b3)								
Pedicularis friderici-augusti	B (b3)		EW		DD				
Pedicularis hoermanniana	B (b3)		LR		LR			R.L. n°9 - 23/02/ 1999	

Pedicularis rostratospicata	B (b3)								
Platanthera algeriensis	B (b3)								Washington Convention
Polygonatum verticillatum	B (b3)		VU	LR	LR				
Potentilla brauneana	B (b3)			VU	LR			R.L. n°9 - 23/02/ 1999	
Pyrola chlorantha	B (b3)			LR					
Pyrus cordata	B (b3)								
Ranunculus magellensis	B (b3)		LR	LR	LR		R.L. n°45 - 11/09/ 1979 and n°66 - 20/06/ 1980	R.L. n°9 - 23/02/ 1999	
Salix breviserrata	B (b3)			LR					
Scabiosa silenifolia	B (b3)		LR	LR					
Scleranthus uncinatus	B (b3)		LR	LR	LR				
Seseli peucedanoides	B (b3)				EW				
Sorbus chamaemespilus	B (b3)		LR	LR	LR			R.L. n°9 - 23/02/ 1999	
Sorbus hybrida	B (b3)								
Sorbus intermedia	B (b3)								
Triglochin palustris	B (b3)			VU					

(continued)

Table 4 (continued)

Taxa	Classes of protection	Red List of Italian Flora (Rossi et al. 2013)	Regional Red List: Latium (Conti et al. 1997)	Regional Red List: Abruzzo (Conti et al. 1997)	Regional Red List: Molise (Conti et al. 1997)	Regional Law (Latium)	Regional Laws (Abruzzo)	Regional Law (Molise)	International Conventions
Valeriana saliunca	B (b3)		LR	LR					
Eriophorum latifolium	B (b3, b4)		EW	EN					
Myosurus minimus	B (b3, B4)			EN	EN			R.L. n°9 - 23/02/ 1999	
Ophrys lacaitae	B (b3, b4, b5)			DD	EN			R.L. n°9 - 23/02/ 1999	Washington Convention
Pinus nigra subsp. nigra var. italica	B (b3, b4, b5)			LR	EN			R.L. n°9 - 23/02/1999	
Betula pendula	B (b3, b5)		EN	LR	VU		R.L. n°45 - 11/09/ 1979 and n°66 - 20/06/1980	R.L. n°9 - 23/02/ 1999	
Leontopodium nivale subsp. nivale	B (b3, b5)		VU	LR			R.L. n°45 - 11/09/ 1979 and n°66 - 20/06/1980		
Pinus mugo subsp. mugo	B (b3, b5)			LR			R.L. n°45 - 11/09/ 1979 and n°66 - 20/06/1980		

Polygala chamaebuxus	B (b3, b5)		LR	LR	LR			R.L. n°9 - 23/02/ 1999	
Ranunculus marsicus	B (b3, b5)			VU	EN			R.L. n°9 - 23/02/ 1999	
Saxifraga glabella	B (b3, b5)		LR	LR	LR			R.L. n°9 - 23/02/ 1999	
Sibbaldia procumbens	B (b3, b5)		LR	LR	LR			R.L. n°9 - 23/02/ 1999	
Silene acaulis subsp. bryoides	B (b3, b5)				LR			R.L. n°9 - 23/02/ 1999	
Streptopus amplexifolius	B (b3, b5)			LR	EN			R.L. n°9 - 23/02/ 1999	
Trollius europaeus subsp. europaeus	B (b3, b5)		VU	LR	EN		R.L. n°45 - 11/09/ 1979 and n°66 - 20/06/1980	R.L. n°9 - 23/02/ 1999	
Euphorbia corallioides	B (b3, b5, b6)				LR	R.L. n°61 - 19/09/ 1996		R.L. n°9 - 23/02/ 1999	
Paeonia officinalis subsp. italica	B (b3, b5, b6)			LR			R.L. n°45 - 11/09/ 1979 and n°66 - 20/06/1980		

(continued)

Table 4 (continued)

Taxa	Classes of protection	Red List of Italian Flora (Rossi et al. 2013)	Regional Red List: Latium (Conti et al. 1997)	Regional Red List: Abruzzo (Conti et al. 1997)	Regional Red List: Molise (Conti et al. 1997)	Regional Law (Latium)	Regional Laws (Abruzzo)	Regional Law (Molise)	International Conventions
Papaver alpinum subsp. alpinum	B (b3, b5, b6)						R.L. n°45 - 11/09/ 1979 and n°66 - 20/06/ 1980		
Phlomis fruticosa	B (b3, b6)			LR					
Vaccinium myrtillus	B (b3, b6)		LR	LR	LR				
Menyanthes trifoliata	B (b4, b5)		EW	EN	CR			R.L. n°9 - 23/02/ 1999	
Tulipa pumila	B (b6)		EW						
Bistorta vivipara	C (2)		VU						
Geranium tuberosum subsp. tuberosum	C (2)		VU	LR	DD				
Geranium versicolor	C (2)		VU						
Alyssum diffusum subsp. diffusum	C (c1)								
Aquilegia dumeticola	C (c1)								
Arenaria bertolonii	C (c1)								

Betonica alopecuros subsp. divulsa	C (c1)								
Brachypodium genuense	C (c1)								
Campanula tanfanii	C (c1)								
Carduus nutans subsp. perspinosus	C (c1)								
Centaurea ambigua subsp. ambigua	C (c1)								
Centaurea ambigua subsp. nigra	C (c1)								
Centaurea nigrescens subsp. neapolitana	C (c1)								
Cerastium scaranoi	C (c1)								
Cerastium tomentosum	C (c1)								
Cirsium lobelii	C (c1)								
Cirsium tenoreanum	C (c1)								
Colchicum neapolitanum	C (c1)								
Dianthus brachycalyx	C (c1)								
Dianthus carthusianorum subsp. tenorei	C (c1)								
Digitalis micrantha	C (c1)								

(continued)

Table 4 (continued)

Taxa	Classes of protection	Red List of Italian Flora (Rossi et al. 2013)	Regional Red List: Latium (Conti et al. 1997)	Regional Red List: Abruzzo (Conti et al. 1997)	Regional Red List: Molise (Conti et al. 1997)	Regional Law (Latium)	Regional Laws (Abruzzo)	Regional Law (Molise)	International Conventions
Drymochloa drymeja subsp. exaltata	C (c1)								
Erysimum pseudorhaeticum	C (c1)								
Festuca alfrediana subsp. ferrariniana	C (c1)								
Festuca violacea subsp. italica	C (c1)								
Galium magellense	C (c1)								
Galium pallidum	C (c1)								
Gentianella columnae	C (c1)								
Helictochloa praetutiana subsp. praetutiana	C (c1)								
Koeleria splendens	C (c1)								
Laserpitium siler subsp. siculum	C (c1)								
Leontodon intermedius	C (c1)								
Luzula sylvatica subsp. sicula	C (c1)								

Melampyrum italicum	C (c1)								
Micromeria graeca subsp. tenuifolia	C (c1)								
Myosotis decumbens subsp. florentina	C (c1)								
Myosotis graui	C (c1)								
Onosma echioides subsp. echioides	C (c1)								
Ornithogalum etruscum subsp. etruscum	C (c1)								
Ornithogalum orthophyllum subsp. orthophyllum	C (c1)								
Pedicularis elegans	C (c1)								
Polygala flavescens	C (c1)								
Potentilla rigoana	C (c1)								
Pulmonaria hirta subsp. apennina	C (c1)								
Ranunculus pollinensis	C (c1)								
Rhinanthus wettsteinii	C (c1)								
Scorzoneroides montana subsp. breviscapa	C (c1)								

(continued)

Table 4 (continued)

Taxa	Classes of protection	Red List of Italian Flora (Rossi et al. 2013)	Regional Red List: Latium (Conti et al. 1997)	Regional Red List: Abruzzo (Conti et al. 1997)	Regional Red List: Molise (Conti et al. 1997)	Regional Law (Latium)	Regional Laws (Abruzzo)	Regional Law (Molise)	International Conventions
Sedum magellense subsp. magellense	C (c1)								
Senecio ovatus subsp. stabianus	C (c1)								
Senecio scopolii subsp. floccosus	C (c1)								
Sesleria nitida	C (c1)								
Silene cattariniana	C (c1)								
Silene roemeri subsp. staminea	C (c1)								
Stachys italica	C (c1)								
Tragopogon porrifolius subsp. eriospermus	C (c1)								
Trifolium pratense subsp. semipurpureum	C (c1)								
Viola eugeniae subsp. eugeniae	C (c1)								
Viola eugeniae subsp. levieri	C (c1)								
Achillea barrelieri subsp. mucronulata	C (c1, c2)		LR						
Achillea tenorei	C (c1, c2)		LR						

Ajuga tenorei	C (c1, c2)			LR					
Asperula cynanchica subsp. neglecta	C (c1, c2)		LR						
Carduus affinis subsp. affinis	C (c1, c2)		LR						
Centaurea ceratophylla subsp. ceratophylla	C (c1, c2)		LR						
Erodium alpinum	C (c1, c2)		VU	LR					
Ornithogalum exscapum	C (c1, c2)		VU						
Oxytropis pilosa subsp. caputoi	C (c1, c2)		LR	LR					
Ranunculus thomasii	C (c1, c2)		LR						
Trisetaria villosa	C (c1, c2)		VU	LR	DD				
Ophrys crabronifera	C (c1, c2, c3, c4)			LR	LR			R.L. n°9 - 23/02/ 1999	Washington Convention
Ophrys tenthredinifera subsp. neglecta	C (c1, c2, c3, c4)			LR	LR			R.L. n°9 - 23/02/ 1999	Washington Convention
Achillea barrelieri subsp. barrelieri	C (c1, c2, c4)				LR			R.L. n°9 - 23/02/ 1999	
Astragalus sirinicus	C (c1, c2, c4)				LR				

(continued)

Table 4 (continued)

Taxa	Classes of protection	Red List of Italian Flora (Rossi et al. 2013)	Regional Red List: Latium (Conti et al. 1997)	Regional Red List: Abruzzo (Conti et al. 1997)	Regional Red List: Molise (Conti et al. 1997)	Regional Law (Latium)	Regional Laws (Abruzzo)	Regional Law (Molise)	International Conventions
								R.L. n°9 - 23/02/ 1999	
Astrantia pauciflora subsp. tenorei	C (c1, c2, c4)		LR	LR	LR			R.L. n°9 - 23/02/ 1999	
Aubrieta columnae subsp. columnae	C (c1, c2, c4)		VU	LR	VU			R.L. n°9 - 23/02/ 1999	
Campanula fragilis subsp. cavolinii	C (c1, c2, c4)		LR	LR	LR			R.L. n°9 - 23/02/ 1999	
Campanula micrantha	C (c1, c2, c4)				LR			R.L. n°9 - 23/02/ 1999	
Coristospermum cuneifolium	C (c1, c2, c4)		LR	LR	LR			R.L. n°9 - 23/02/ 1999	
Cymbalaria pallida	C (c1, c2, c4)		LR	LR	LR			R.L. n°9 - 23/02/ 1999	
Erysimum majellense	C (c1, c2, c4)				LR			R.L. n°9 - 23/02/ 1999	

Jacobaea alpina subsp. samnitum	C (c1, c2, c4)				LR			R.L. n°9 - 23/02/ 1999	
Leucanthemum coronopifolium subsp. tenuifolium	C (c1, c2, c4)			LR	LR			R.L. n°9 - 23/02/ 1999	
Melampyrum variegatum	C (c1, c2, c4)			LR	LR			R.L. n°9 - 23/02/ 1999	
Minuartia graminifolia subsp. rosanoi	C (c1, c2, c4)		LR		LR			R.L. n°9 - 23/02/ 1999	
Noccaea stylosa	C (c1, c2, c4)			LR	LR			R.L. n°9 - 23/02/ 1999	
Ranunculus apenninus	C (c1, c2, c4)				LR			R.L. n°9 - 23/02/ 1999	
Saxifraga exarata subsp. ampullacea	C (c1, c2, c4)			LR	LR			R.L. n°9 - 23/02/ 1999	
Saxifraga porophylla subsp. porophylla	C (c1, c2, c4)			LR	LR		R.L. n°9 - 23/02/ 1999		
Sempervivum riccii	C (c1, c2, c4)			LR	LR			R.L. n°9 - 23/02/ 1999	
Stipa dasyvaginata subsp. apenninicola	C (c1, c2, c4)				LR			R.L. n°9 - 23/02/ 1999	

(continued)

Table 4 (continued)

Taxa	Classes of protection	Red List of Italian Flora (Rossi et al. 2013)	Regional Red List: Latium (Conti et al. 1997)	Regional Red List: Abruzzo (Conti et al. 1997)	Regional Red List: Molise (Conti et al. 1997)	Regional Law (Latium)	Regional Laws (Abruzzo)	Regional Law (Molise)	International Conventions
Taraxacum glaciale	C (c1, c2, c4)		LR	LR	VU			R.L. n°9 - 23/02/ 1999	
Verbascum niveum subsp. garganicum	C (c1, c2, c4)		LR	VU		R.L. n°61 - 19/09/ 1976	R.L. n°45 - 11/09/ 1979 and n°66 - 20/06/ 1980		
Epipactis helleborine subsp. latina	C (c1, c3)								Washington Convention
Epipactis meridionalis	C (c1, c3)								Washington Convention
Ophrys appennina	C (c1, c3)								Washington Convention
Ophrys ausonia	C (c1, c3)								Washington Convention
Ophrys lucana	C (c1, c3)								Washington Convention
Ophrys promontorii	C (c1, c3)								Washington Convention
									Washington Convention

Cynoglossum apenninum	C (c1, c4)					R.L. n°61 - 19/09/ 1992			
Cynoglossum magellense	C (c1, c4)					R.L. n°61 - 19/09/ 1993			
Hypochaeris robertia	C (c1, c4)					R.L. n°61 - 19/09/ 2000			
Linaria purpurea	C (c1, c4)					R.L. n°61 - 19/09/ 2003			
Lomelosia crenata subsp. pseudisetensis	C (c1, c4)				DD			R.L. n°9 - 23/02/ 1999	
Aconitum lycoctonum	C (c2)		LR						
Adonis annua	C (c2)			VU					
Agrostemma githago	C (c2)			VU					
Anemonastrum narcissiflorum subsp. narcissiflorum	C (c2)		LR						
Arenaria grandiflora subsp. grandiflora	C (c2)		LR						
Asplenium adiantum-nigrum subsp. adiantum-nigrum	C (c2)		LR						

(continued)

Table 4 (continued)

Taxa	Classes of protection	Red List of Italian Flora (Rossi et al. 2013)	Regional Red List: Latium (Conti et al. 1997)	Regional Red List: Abruzzo (Conti et al. 1997)	Regional Red List: Molise (Conti et al. 1997)	Regional Law (Latium)	Regional Laws (Abruzzo)	Regional Law (Molise)	International Conventions
Asplenium ceterach subsp. ceterach	C (c2)		LR						
Asplenium lepidum subsp. lepidum	C (c2)		EW	LR					
Aster alpinus subsp. alpinus	C (c2)		VU	LR					
Barbarea stricta	C (c2)			VU					
Berberis vulgaris subsp. vulgaris	C (c2)				LR				
Blysmus compressus	C (c2)		LR						
Buglossoides incrassata	C (c2)		LR						
Bupleurum rotundifolium	C (c2)		VU						
Camelina sativa	C (c2)			VU					
Campanula foliosa	C (c2)		LR						
Carex distachya	C (c2)			VU					
Carex elata subsp. elata	C (c2)		LR	VU					
Carex flava	C (c2)		VU	VU					
Carex nigra subsp. nigra	C (c2)		LR	VU					
Carex olbiensis	C (c2)				LR				

Carex panicea	C (c2)		VU	VU					
Carex punctata	C (c2)		LR	VU					
Carex remota	C (c2)			VU					
Carex rostrata	C (c2)		VU	VU					
Carex tomentosa	C (c2)			VU					
Carex umbrosa subsp. umbrosa	C (c2)		LR	VU					
Catabrosa aquatica	C (c2)			VU					
Cerastium siculum	C (c2)		VU						
Ceratocephala falcata	C (c2)		EW	VU					
Cerinthe minor subsp. auriculata	C (c2)		LR	LR					
Chamaeiris foetidissima	C (c2)			VU					
Cirsium creticum subsp. triumfetti	C (c2)			VU					
Conringia orientalis	C (c2)		LR						
Coronilla repanda subsp. repanda	C (c2)		LR						
Crocus biflorus	C (c2)				LR			R.L. n°9 - 23/02/ 1999	
Crucianella angustifolia	C (c2)		LR	LR					
Cytisus decumbens	C (c2)			LR					

(continued)

Table 4 (continued)

Taxa	Classes of protection	Red List of Italian Flora (Rossi et al. 2013)	Regional Red List: Latium (Conti et al. 1997)	Regional Red List: Abruzzo (Conti et al. 1997)	Regional Red List: Molise (Conti et al. 1997)	Regional Law (Latium)	Regional Laws (Abruzzo)	Regional Law (Molise)	International Conventions
Delphinium pubescens	C (c2)			VU					
Descurainia sophia	C (c2)		LR		EW				
Eleocharis quinqueflora	C (c2)			LR					
Ephedra major subsp. major	C (c2)			LR					
Equisetum hyemale	C (c2)		LR						
Euphrasia illyrica	C (c2)		LR						
Euphrasia italica	C (c2)			LR					
Falcaria vulgaris	C (c2)			LR					
Festuca bosniaca	C (c2)			LR					
Fibigia clypeata	C (c2)		VU						
Gagea bohemica	C (c2)		LR	LR					
Gagea fragifera	C (c2)		LR		LR			R.L. n°9 - 23/02/ 1999	
Gagea minima	C (c2)		LR						
Gagea pratensis	C (c2)		LR						
Galium rotundifolium subsp. rotundifolium	C (c2)			LR					
Galium verticillatum	C (c2)			LR					
Gentiana utriculosa	C (c2)		VU						

Gentianopsis ciliata subsp. ciliata	C (c2)		VU						
Gymnocarpium robertianum	C (c2)		LR	LR	DD				
Helianthemum nummulariumsubsp. nummularium	C (c2)		LR						
Hippocrepis glauca	C (c2)		LR						
Hyoseris scabra	C (c2)			LR					
Hypericum androsaemum	C (c2)			VU					
Iberis saxatilis. subsp. saxatilis	C (c2)		VU						
Inula hirta.	C (c2)		LR						
Lamium flexuosum	C (c2)		VU						
Lamium galeobdolon subsp. galeobdolon	C (c2)			LR	VU				
Lathyrus nissolia	C (c2)		VU						
Lens nigricans	C (c2)		EW	LR					
Lysimachia nemorum	C (c2)		VU						
Melica nutans	C (c2)			LR	LR				
Moneses uniflora	C (c2)			LR					
Myagrum perfoliatum	C (c2)		VU						
Myosotis incrassata	C (c2)		VU						

(continued)

Table 4 (continued)

Taxa	Classes of protection	Red List of Italian Flora (Rossi et al. 2013)	Regional Red List: Latium (Conti et al. 1997)	Regional Red List: Abruzzo (Conti et al. 1997)	Regional Red List: Molise (Conti et al. 1997)	Regional Law (Latium)	Regional Laws (Abruzzo)	Regional Law (Molise)	International Conventions
Neslia paniculata subsp. thracica	C (c2)		LR						
Oenanthe silaifolia	C (c2)		LR	DD					
Onobrychis alba subsp. alba	C (c2)		LR						
Oreoselinum nigrum	C (c2)		LR						
Ornithogalum comosum	C (c2)		VU						
Orobanche purpurea	C (c2)		LR						
Orobanche teucrii	C (c2)		LR						
Papaver apulum	C (c2)		LR	VU					
Paris quadrifolia	C (c2)		LR						
Persicaria hydropiper	C (c2)			VU					
Polypodium vulgare	C (c2)		LR						
Potamogeton berchtoldii	C (c2)			VU	LR				
Potamogeton lucens	C (c2)			VU					
Potamogeton polygonifolius	C (c2)		LR	VU					
Potamogeton trichoides	C (c2)		LR	VU					

Potentilla erecta	C (c2)		LR						
Potentilla inclinata	C (c2)			LR					
Pyrola minor	C (c2)		VU						
Ranunculus parviflorus	C (c2)		LR		LR				
Ranunculus polyanthemoides	C (c2)			LR					
Ribes alpinum	C (c2)			LR					
Salvia argentea	C (c2)		VU						
Saponaria bellidifolia	C (c2)		LR	LR					
Schedonorus giganteus	C (c2)			VU					
Scorzonera austriaca	C (c2)		LR	LR					
Scorzoneroides autumnalis	C (c2)		LR						
Scutellaria alpina subsp. alpina	C (c2)		LR	LR					
Seseli tommasinii	C (c2)		LR						
Seseli tortuosum subsp. tortuosum	C (c2)			LR					
Silene ciliata subsp. graefferi	C (c2)		LR						
Stellaria aquatica	C (c2)			VU					
Stellaria graminea	C (c2)		LR	VU					
Sternbergia colchiciflora	C (c2)		VU	LR					
Thesium parnassi	C (c2)		LR						

(continued)

Table 4 (continued)

Taxa	Classes of protection	Red List of Italian Flora (Rossi et al. 2013)	Regional Red List: Latium (Conti et al. 1997)	Regional Red List: Abruzzo (Conti et al. 1997)	Regional Red List: Molise (Conti et al. 1997)	Regional Law (Latium)	Regional Laws (Abruzzo)	Regional Law (Molise)	International Conventions
Trifolium hybridum subsp. elegans	C (c2)		LR						
Trifolium hybridum subsp. hybridum	C (c2)			LR					
Trifolium rubens	C (c2)		VU						
Trigonella gladiata	C (c2)			LR					
Valerianella echinata	C (c2)		LR						
Valerianella pumila	C (c2)		LR						
Veratrum album	C (c2)				LR				
Verbascum alpinum	C (c2)		LR	LR					
Veronica agrestis	C (c2)		LR		DD				
Veronica austriaca	C (c2)		LR						
Veronica praecox	C (c2)		LR						
Veronica verna subsp. verna	C (c2)		LR	LR	LR				
Viburnum opulus	C (c2)		LR	VU					
Vicia disperma	C (c2)		VU						
Viola kitaibeliana	C (c2)		LR						
Viscum album subsp. album	C (c2)			LR					
Orchis militaris	C (c2, c3)			LR					Washington Convention

Orchis simia	C (c2, c3)				LR			R.L. n°9 - 23/02/ 1999	Washington Convention
Serapias lingua	C (c2, c3)			LR					Washington Convention
Anacamptis laxiflora	C (c2, c3, c4)			EN	VU			R.L. n°9 - 23/02/ 1999	Washington Convention
Corallorhiza trifida	C (c2, c3, c4)				LR			R.L. n°9 - 23/02/ 1999	Washington Convention
Epipactis muelleri	C (c2, c3, c4)				LR			R.L. n°9 - 23/02/ 1999	Washington Convention
Epipogium aphyllum	C (c2, c3, c4)		LR	VU	LR			R.L. n°9 - 23/02/ 1999	Washington Convention
Gentiana lutea subsp. lutea	C (c2, c3, c4)		VU	VU	VU		R.L. n°45 - 11/09/ 1979 and n°66 - 20/06/ 1980	R.L. n°9 - 23/02/ 1999	Habitat Directive 92/43/CEE and Washington Convention
Nigritella widderi	C (c2, c3, c4)		LR	LR	LR		R.L. n°45 - 11/09/ 1979 and n°66 - 20/06/ 1980	R.L. n°9 - 23/02/ 1999	Washington Convention

(continued)

Table 4 (continued)

Taxa	Classes of protection	Red List of Italian Flora (Rossi et al. 2013)	Regional Red List: Latium (Conti et al. 1997)	Regional Red List: Abruzzo (Conti et al. 1997)	Regional Red List: Molise (Conti et al. 1997)	Regional Law (Latium)	Regional Laws (Abruzzo)	Regional Law (Molise)	International Conventions
Ophrys insectifera	C (c2, c3, c4)			LR	LR			R.L. n°9 - 23/02/1999	Washington Convention
Orchis spitzelii	C (c2, c3, c4)		LR	LR					Washington Convention
Pseudorchis albida	C (c2, c3, c4)		LR	LR	EN			R.L. n°9 - 23/02/1999	Washington Convention
Serapias parviflora	C (c2, c3, c4)				LR			R.L. n°9 - 23/02/1999	Washington Convention
Serratula tinctoria subsp. tinctoria	C (c2, c3, c4)			VU	LR			R.L. n°9 - 23/02/1999	
Typha minima	C (c2, c3, c4)		VU	VU	LR			R.L. n°9 - 23/02/1999	Berne Convention
Dactylorhiza incarnata subsp. incarnata	C (c2, c3, c4, c5)		VU	VU	EN			R.L. n°9 - 23/02/1999	Washington Convention
Epipactis palustris	C (c2, c3, c4, c5)			VU	CR			R.L. n°9 - 23/02/1999	Washington Convention

Adonis flammea subsp. flammea	C (c2, c4)		LR	VU	LR			R.L. n°9 - 23/02/ 1999	
Allium cupanii subsp. cupanii	C (c2, c4)		LR		LR			R.L. n°9 - 23/02/ 1999	
Allium flavum subsp. flavum	C (c2, c4)			LR	LR			R.L. n°9 - 23/02/ 1999	
Allium saxatile subsp. tergestinum	C (c2, c4)		LR	LR	VU			R.L. n°9 - 23/02/ 1999	
Alopecurus aequalis	C (c2, c4)		LR	VU	VU			R.L. n°9 - 23/02/ 1999	
Anemonoides ranunculoides	C (c2, c4)				LR			R.L. n°9 - 23/02/ 1999	
Anthemis cretica subsp. columnae	C (c2, c4)		LR		LR			R.L. n°9 - 23/02/ 1999	
Anthriscus nitida	C (c2, c4)		LR		LR			R.L. n°9 - 23/02/ 1999	
Arabis auriculata	C (c2, c4)		LR		LR			R.L. n°9 - 23/02/ 1999	
Arabis collina subsp. rosea	C (c2, c4)				LR			R.L. n°9 - 23/02/ 1999	

(continued)

Table 4 (continued)

Taxa	Classes of protection	Red List of Italian Flora (Rossi et al. 2013)	Regional Red List: Latium (Conti et al. 1997)	Regional Red List: Abruzzo (Conti et al. 1997)	Regional Red List: Molise (Conti et al. 1997)	Regional Law (Latium)	Regional Laws (Abruzzo)	Regional Law (Molise)	International Conventions
Arabis surculosa	C (c2, c4)		LR		LR			R.L. n°9 - 23/02/ 1999	
Arum cylindraceum	C (c2, c4)			LR	LR			R.L. n°9 - 23/02/ 1999	
Asphodeline lutea	C (c2, c4)		LR		LR	R.L. n°61 - 19/09/ 1975		R.L. n°9 - 23/02/ 1999	
Astragalus vesicarius subsp. vesicarius	C (c2, c4)		LR	LR	LR			R.L. n°9 - 23/02/ 1999	
Astrantia major subsp. involucrata	C (c2, c4)				LR			R.L. n°9 - 23/02/ 1999	
Athamanta sicula	C (c2, c4)			LR	VU	R.L. n°61 - 19/09/ 1984		R.L. n°9 - 23/02/ 1999	
Bellidiastrum michelii	C (c2, c4)				LR			R.L. n°9 - 23/02/ 1999	
Biarum tenuifolium subsp. tenuifolium	C (c2, c4)			LR		R.L. n°61 - 19/09/ 1986			

Brassica gravinae	C (c2, c4)		LR	LR	LR			R.L. n°9 - 23/02/ 1999	
Campanula bononiensis	C (c2, c4)		VU	LR	LR			R.L. n°9 - 23/02/ 1999	
Campanula cochleariifolia	C (c2, c4)				LR			R.L. n°9 - 23/02/ 1999	
Campanula latifolia	C (c2, c4)		VU		LR			R.L. n°9 - 23/02/ 1999	
Carduus chrysacanthus	C (c2, c4)				LR			R.L. n°9 - 23/02/ 1999	
Carex acuta	C (c2, c4)			VU				R.L. n°9 - 23/02/ 1999	
Carex acutiformis	C (c2, c4)			VU	VU			R.L. n°9 - 23/02/ 1999	
Carex digitata	C (c2, c4)		LR		LR			R.L. n°9 - 23/02/ 1999	
Carex divisa.	C (c2, c4)				LR			R.L. n°9 - 23/02/ 1999	
Carex ornithopoda	C (c2, c4)			LR	LR			R.L. n°9 - 23/02/ 1999	

(continued)

Table 4 (continued)

Taxa	Classes of protection	Red List of Italian Flora (Rossi et al. 2013)	Regional Red List: Latium (Conti et al. 1997)	Regional Red List: Abruzzo (Conti et al. 1997)	Regional Red List: Molise (Conti et al. 1997)	Regional Law (Latium)	Regional Laws (Abruzzo)	Regional Law (Molise)	International Conventions
Carex pilosa	C (c2, c4)			LR	LR			R.L. n°9 - 23/02/1999	
Carex riparia	C (c2, c4)			VU	LR			R.L. n°9 - 23/02/1999	
Carex vesicaria	C (c2, c4)		VU	VU	LR			R.L. n°9 - 23/02/1999	
Carlina acanthifolia subsp. acanthifolia	C (c2, c4)			LR	LR		R.L. n°45 - 11/09/1979 and n°66 - 20/06/1980	R.L. n°9 - 23/02/1999	
Cerastium sylvaticum	C (c2, c4)				LR			R.L. n°9 - 23/02/1999	
Cirsium acaulon subsp. acaulon	C (c2, c4)		LR		LR			R.L. n°9 - 23/02/1999	
Cirsium palustre	C (c2, c4)			VU	VU			R.L. n°9 - 23/02/1999	

Clypeola jonthlaspi subsp. jonthlaspi	C (c2, c4)		LR		VU			R.L. n°9 - 23/02/ 1999	
Convallaria majalis	C (c2, c4)		VU	LR	VU			R.L. n°9 - 23/02/ 1999	
Corydalis pumila	C (c2, c4)			LR	LR			R.L. n°9 - 23/02/ 1999	
Crepis biennis	C (c2, c4)		LR		LR			R.L. n°9 - 23/02/ 1999	
Crypsis alopecuroides	C (c2, c4)			EW	LR			R.L. n°9 - 23/02/ 1999	
Cucubalus baccifer	C (c2, c4)			VU	LR			R.L. n°9 - 23/02/ 1999	
Cynosurus effusus	C (c2, c4)				LR			R.L. n°9 - 23/02/ 1999	
Daphne alpina subsp. alpina	C (c2, c4)		LR		LR			R.L. n°9 - 23/02/ 1999	
Daphne sericea	C (c2, c4)			LR	VU	R.L. n°61 - 19/09/ 1995		R.L. n°9 - 23/02/ 1999	
Dianthus ciliatus subsp. ciliatus	C (c2, c4)				LR			R.L. n°9 - 23/02/ 1999	

(continued)

Table 4 (continued)

Taxa	Classes of protection	Red List of Italian Flora (Rossi et al. 2013)	Regional Red List: Latium (Conti et al. 1997)	Regional Red List: Abruzzo (Conti et al. 1997)	Regional Red List: Molise (Conti et al. 1997)	Regional Law (Latium)	Regional Laws (Abruzzo)	Regional Law (Molise)	International Conventions
Dictamnus albus	C (c2, c4)		VU	VU	DD		R.L. n°45 - 11/09/ 1979 and n°66 - 20/06/ 1980		
Dryopteris villarii subsp. villarii	C (c2, c4)		LR		LR			R.L. n°9 - 23/02/ 1999	
Equisetum fluviatile	C (c2, c4)		EW	VU	LR			R.L. n°9 - 23/02/ 1999	
Erica multiflora subsp. multiflora	C (c2, c4)			LR	LR			R.L. n°9 - 23/02/ 1999	
Euonymus verrucosus	C (c2, c4)		LR		LR			R.L. n°9 - 23/02/ 1999	
Euphrasia liburnica	C (c2, c4)				LR			R.L. n°9 - 23/02/ 1999	
Ferula glauca	C (c2, c4)				LR	R.L. n°61 - 19/09/ 1997		R.L. n°9 - 23/02/ 1999	

Filipendula ulmaria	C (c2, c4)		VU	VU	LR			R.L. n°9 - 23/02/ 1999	
Fraxinus angustifolia subsp. oxycarpa (	C (c2, c4)			VU	VU			R.L. n°9 - 23/02/ 1999	
Fumana ericifolia	C (c2, c4)				LR			R.L. n°9 - 23/02/ 1999	
Gagea granatellii	C (c2, c4)				LR			R.L. n°9 - 23/02/ 1999	
Genista sagittalis	C (c2, c4)		LR	LR	LR			R.L. n°9 - 23/02/ 1999	
Gentiana dinarica	C (c2, c4)		VU		LR		R.L. n°45 - 11/09/ 1979 and n°66 - 20/06/ 1980	R.L. n°9 - 23/02/ 1999	
Geranium macrorrhizum	C (c2, c4)		LR	LR	LR			R.L. n°9 - 23/02/ 1999	
Geranium reflexum	C (c2, c4)				LR			R.L. n°9 - 23/02/ 1999	
Geropogon hybridus	C (c2, c4)			LR		R.L. n°61 - 19/09/ 1998			

(continued)

Table 4 (continued)

Taxa	Classes of protection	Red List of Italian Flora (Rossi et al. 2013)	Regional Red List: Latium (Conti et al. 1997)	Regional Red List: Abruzzo (Conti et al. 1997)	Regional Red List: Molise (Conti et al. 1997)	Regional Law (Latium)	Regional Laws (Abruzzo)	Regional Law (Molise)	International Conventions
Gnaphalium uliginosum	C (c2, c4)		LR					R.L. n°9 - 23/02/ 1999	
Groenlandia densa	C (c2, c4)			VU	VU			R.L. n°9 - 23/02/ 1999	
Hesperis matronalis subsp. matronalis	C (c2, c4)				LR	R.L. n°61 - 19/09/ 1974		R.L. n°9 - 23/02/ 1999	
Holosteum umbellatum subsp. umbellatum	C (c2, c4)		EW		CR			R.L. n°9 - 23/02/ 1999	
Hypericum richeri subsp. richeri	C (c2, c4)				LR			R.L. n°9 - 23/02/ 1999	
Impatiens noli-tangere	C (c2, c4)		LR		LR			R.L. n°9 - 23/02/ 1999	
Inula helenium	C (c2, c4)			VU	VU			R.L. n°9 - 23/02/ 1999	
Isolepis cernua	C (c2, c4)			LR	LR			R.L. n°9 - 23/02/ 1999	

Jurinea mollis subsp. mollis	C (c2, c4)		LR		LR			R.L. n°9 - 23/02/ 1999	
Kernera saxatilis subsp. saxatilis	C (c2, c4)		LR		LR			R.L. n°9 - 23/02/ 1999	
Laburnum alpinum	C (c2, c4)			LR	LR			R.L. n°9 - 23/02/ 1999	
Lathyrus pannonicus	C (c2, c4)		VU	VU	LR			R.L. n°9 - 23/02/ 1999	
Lilium bulbiferum subsp. croceum	C (c2, c4)		VU	LR	LR	R.L. n°61 - 19/09/ 1977	R.L. n°45 - 11/09/ 1979 and n°66 - 20/06/ 1980	R.L. n°9 - 23/02/ 1999	
Lilium martagon	C (c2, c4)		VU	LR	LR		R.L. n°45 - 11/09/ 1979 and n°66 - 20/06/ 1980	R.L. n°9 - 23/02/ 1999	
Limniris pseudacorus	C (c2, c4)			VU	VU			R.L. n°9 - 23/02/ 1999	

(continued)

Table 4 (continued)

Taxa	Classes of protection	Red List of Italian Flora (Rossi et al. 2013)	Regional Red List: Latium (Conti et al. 1997)	Regional Red List: Abruzzo (Conti et al. 1997)	Regional Red List: Molise (Conti et al. 1997)	Regional Law (Latium)	Regional Laws (Abruzzo)	Regional Law (Molise)	International Conventions
Linum capitatum subsp. serrulatum	C (c2, c4)				LR			R.L. n°9 - 23/02/ 1999	
Lomelosia graminifolia subsp. graminifolia	C (c2, c4)				LR			R.L. n°9 - 23/02/ 1999	
Medicago disciformis	C (c2, c4)		VU		LR			R.L. n°9 - 23/02/ 1999	
Mercurialis ovata	C (c2, c4)		LR	LR	LR			R.L. n°9 - 23/02/ 1999	
Myosotis laxa subsp. caespitosa	C (c2, c4)		LR	VU	VU			R.L. n°9 - 23/02/ 1999	
Myosotis nemorosa	C (c2, c4)		LR		VU			R.L. n°9 - 23/02/ 1999	
Myosotis scorpioides subsp. scorpioides	C (c2, c4)			VU	VU			R.L. n°9 - 23/02/ 1999	
Narcissus poëticus	C (c2, c4)		VU		LR	R.L. n°61 - 19/09/ 1978		R.L. n°9 - 23/02/ 1999	

Oenanthe fistulosa	C (c2, c4)			VU	VU			R.L. n°9 - 23/02/ 1999	
Oxytropis campestris	C (c2, c4)		LR		LR			R.L. n°9 - 23/02/ 1999	
Phalaroides arundinacea subsp. arundinacea	C (c2, c4)			VU	VU			R.L. n°9 - 23/02/ 1999	
Piptatherum virescens	C (c2, c4)				LR	R.L. n°61 - 19/09/ 2004		R.L. n°9 - 23/02/ 1999	
Plantago maritima subsp. serpentina	C (c2, c4)				LR			R.L. n°9 - 23/02/ 1999	
Polycnemum arvense	C (c2, c4)				LR			R.L. n°9 - 23/02/ 1999	
Polygala monspeliaca	C (c2, c4)			LR	LR			R.L. n°9 - 23/02/ 1999	
Potentilla apennina subsp. apennina	C (c2, c4)				LR			R.L. n°9 - 23/02/ 1999	
Primula auricula	C (c2, c4)				LR		R.L. n°45 - 11/09/ 1979 and n°66 - 20/06/ 1980	R.L. n°9 - 23/02/ 1999	

(continued)

Table 4 (continued)

Taxa	Classes of protection	Red List of Italian Flora (Rossi et al. 2013)	Regional Red List: Latium (Conti et al. 1997)	Regional Red List: Abruzzo (Conti et al. 1997)	Regional Red List: Molise (Conti et al. 1997)	Regional Law (Latium)	Regional Laws (Abruzzo)	Regional Law (Molise)	International Conventions
Quercus robur subsp. robur	C (c2, c4)			VU	VU			R.L. n°9 - 23/02/ 1999	
Ranunculus acris subsp. acris	C (c2, c4)				VU			R.L. n°9 - 23/02/ 1999	
Ranunculus nemorosus	C (c2, c4)		LR	VU	LR			R.L. n°9 - 23/02/ 1999	
Rosa montana	C (c2, c4)			LR	LR			R.L. n°9 - 23/02/ 1999	
Rosa spinosissima	C (c2, c4)				LR			R.L. n°9 - 23/02/ 1999	
Rubus saxatilis	C (c2, c4)				LR			R.L. n°9 - 23/02/ 1999	
Ruscus hypoglossum	C (c2, c4)				LR			R.L. n°9 - 23/02/ 1999	
Salix cinerea	C (c2, c4)			VU	LR			R.L. n°9 - 23/02/ 1999	

Salix retusa	C (c2, c4)		LR		LR			R.L. n°9 - 23/02/ 1999	
Saxifraga caesia	C (c2, c4)		LR		LR			R.L. n°9 - 23/02/ 1999	
Saxifraga callosa subsp. callosa	C (c2, c4)				LR			R.L. n°9 - 23/02/ 1999	
Scutellaria altissima	C (c2, c4)		VU	LR	LR			R.L. n°9 - 23/02/ 1999	
Scutellaria galericulata	C (c2, c4)		LR	VU	VU			R.L. n°9 - 23/02/ 1999	
Silene catholica	C (c2, c4)			LR	LR	R.L. n°61 - 19/09/ 2007		R.L. n°9 - 23/02/ 1999	
Silene saxifraga	C (c2, c4)		LR		LR			R.L. n°9 - 23/02/ 1999	
Soldanella alpina subsp. alpina	C (c2, c4)		VU		LR		R.L. n°45 - 11/09/ 1979 and n°66 - 20/06/ 1980	R.L. n°9 - 23/02/ 1999	

(continued)

Table 4 (continued)

Taxa	Classes of protection	Red List of Italian Flora (Rossi et al. 2013)	Regional Red List: Latium (Conti et al. 1997)	Regional Red List: Abruzzo (Conti et al. 1997)	Regional Red List: Molise (Conti et al. 1997)	Regional Law (Latium)	Regional Laws (Abruzzo)	Regional Law (Molise)	International Conventions
Staphylea pinnata	C (c2, c4)				LR	R.L. n°61 - 19/09/ 2008		R.L. n°9 - 23/02/ 1999	
Sternbergia lutea	C (c2, c4)		VU	LR	LR	R.L. n°61 - 19/09/ 1979		R.L. n°9 - 23/02/ 1999	
Tephroseris integrifolia (subsp. integrifolia	C (c2, c4)				LR			R.L. n°9 - 23/02/ 1999	
Thalictrum simplex subsp. simplex	C (c2, c4)			VU	LR			R.L. n°9 - 23/02/ 1999	
Trifolium aureum subsp. aureum	C (c2, c4)				LR			R.L. n°9 - 23/02/ 1999	
Trifolium dubium	C (c2, c4)			VU	LR			R.L. n°9 - 23/02/ 1999	
Trifolium phleoides	C (c2, c4)		VU		LR			R.L. n°9 - 23/02/ 1999	
Trigonella monspeliaca	C (c2, c4)		VU	LR	LR			R.L. n°9 - 23/02/ 1999	

Verbascum chaixii subsp. chaixii	C (c2, c4)		VU		LR			R.L. n°9 - 23/02/ 1999	
Verbascum samniticum	C (c2, c4)				LR			R.L. n°9 - 23/02/ 1999	
Veronica barrelieri subsp. barrelieri	C (c2, c4)		LR	LR	LR			R.L. n°9 - 23/02/ 1999	
Veronica prostrata subsp. prostrata	C (c2, c4)		LR		LR			R.L. n°9 - 23/02/ 1999	
Veronica scutellata	C (c2, c4)		VU	VU	LR			R.L. n°9 - 23/02/ 1999	
Alyssoides utriculata	C (c2, c4, c5)		LR	LR	CR			R.L. n°9 - 23/02/ 1999	
Asarum europaeum	C (c2, c4, c5)		LR	LR	EN			R.L. n°9 - 23/02/ 1999	
Caltha palustris	C (c2, c4, c5)			VU	EN			R.L. n°9 - 23/02/ 1999	
Carex liparocarpos subsp. liparocarpos	C (c2, c4, c5)		LR	LR	CR			R.L. n°9 - 23/02/ 1999	
Carex paniculata subsp. paniculata	C (c2, c4, c5)		CR	VU	CR			R.L. n°9 - 23/02/ 1999	

(continued)

Table 4 (continued)

Taxa	Classes of protection	Red List of Italian Flora (Rossi et al. 2013)	Regional Red List: Latium (Conti et al. 1997)	Regional Red List: Abruzzo (Conti et al. 1997)	Regional Red List: Molise (Conti et al. 1997)	Regional Law (Latium)	Regional Laws (Abruzzo)	Regional Law (Molise)	International Conventions
Eleocharis uniglumis subsp. uniglumis	C (c2, c4, c5)		CR	VU	VU			R.L. n°9 - 23/02/ 1999	
Fritillaria montana	C (c2, c4, c5)		VU	LR	CR			R.L. n°9 - 23/02/ 1999	
Linaria simplex	C (c2, c4, c5)		VU		CR			R.L. n°9 - 23/02/ 1999	
Ophioglossum vulgatum subsp. vulgatum	C (c2, c4, c5)		VU	VU	EN			R.L. n°9 - 23/02/ 1999	
Ranunculus flammula	C (c2, c4, c5)		VU	EW	EN			R.L. n°9 - 23/02/ 1999	
Arabis pauciflora	C (c2, c5)		CR	LR					
Epilobium palustre	C (c2, c5)			VU	CR				
Myosotis stricta	C (c2, c5)		EN	LR	LR				
Anacamptis coriophora	C (c3)								Washington Convention
Anacamptis morio	C (c3)								Washington Convention
Anacamptis pyramidalis	C (c3)								Washington Convention

Arctostaphylos uva-ursi	C (c3)								Washington Convention
Cephalanthera damasonium	C (c3)								Washington Convention
Cephalanthera longifolia	C (c3)								Washington Convention
Cephalanthera rubra	C (c3)								Washington Convention
Coeloglossum viride	C (c3)								Washington Convention
Cyclamen hederifolium subsp. hederifolium	C (c3)								Washington Convention
Cyclamen repandum subsp. repandum	C (c3)								Washington Convention
Dactylorhiza maculata subsp. fuchsii	C (c3)								Washington Convention
Dactylorhiza maculata subsp. saccifera	C (c3)								Washington Convention
Dactylorhiza sambucina	C (c3)								Washington Convention
Epipactis atrorubens	C (c3)								Washington Convention
Epipactis helleborine subsp. helleborine	C (c3)								Washington Convention
Epipactis leptochila	C (c3)								Washington Convention

(continued)

Table 4 (continued)

Taxa	Classes of protection	Red List of Italian Flora (Rossi et al. 2013)	Regional Red List: Latium (Conti et al. 1997)	Regional Red List: Abruzzo (Conti et al. 1997)	Regional Red List: Molise (Conti et al. 1997)	Regional Law (Latium)	Regional Laws (Abruzzo)	Regional Law (Molise)	International Conventions
Epipactis microphylla	C (c3)								Washington Convention
Epipactis purpurata	C (c3)			LR					Washington Convention
Gymnadenia conopsea	C (c3)								Washington Convention
Himantoglossum adriaticum	C (c3)								Habitat Directive 92/43/CEE and Washington Convention
Limodorum abortivum	C (c3)								Washington Convention
Listera ovata	C (c3)								Washington Convention
Neotinea tridentata	C (c3)								Washington Convention
Neotinea ustulata	C (c3)								Washington Convention
Neottia nidus- avis	C (c3)								Washington Convention
Ophrys apifera	C (c3)								Washington Convention

Ophrys bertolonii subsp. bertolonii	C (c3)								
Ophrys dinarica	C (c3)								Washington Convention
Ophrys exaltata subsp. archipelagi	C (c3)								Washington Convention
Ophrys funerea	C (c3)								
Ophrys gracilis	C (c3)								Washington Convention
Ophrys incubacea	C (c3)								Washington Convention
Ophrys passionis subsp. passionis	C (c3)								Washington Convention
Ophrys tetraloniae	C (c3)								Washington Convention
Orchis anthropophora	C (c3)								Washington Convention
Orchis italica	C (c3)								Washington Convention
Orchis mascula subsp. mascula	C (c3)								Washington Convention
Orchis mascula subsp. speciosa	C (c3)								Washington Convention
Orchis pauciflora	C (c3)								Washington Convention
Orchis provincialis	C (c3)								Washington Convention,

(continued)

Table 4 (continued)

Taxa	Classes of protection	Red List of Italian Flora (Rossi et al. 2013)	Regional Red List: Latium (Conti et al. 1997)	Regional Red List: Abruzzo (Conti et al. 1997)	Regional Red List: Molise (Conti et al. 1997)	Regional Law (Latium)	Regional Laws (Abruzzo)	Regional Law (Molise)	International Conventions
									Berne Convention
Platanthera bifolia	C (c3)								Washington Convention
Platanthera chlorantha	C (c3)								Washington Convention
Serapias bergonii	C (c3)								Washington Convention
Serapias vomeracea	C (c3)								Washington Convention
Spiranthes spiralis	C (c3)								Washington Convention
Epipactis persica subsp. gracilisi	C (c3, c4)							R.L. n°9 - 23/02/ 1999	Washington Convention
Galanthus nivalis	C (c3, c4)					R.L. n°61 - 19/09/ 1981			Habitat Directive 92/43/CEE and Washington Convention
Orchis pallens	C (c3, c4)					R.L. n°61 - 19/09/ 1980			Washington Convention

Anemone apennina subsp. apennina	C (c4)						R.L. n°45 - 11/09/ 1979 and n°66 - 20/06/ 1980		
Anthyllis montana subsp. jacquinii	C (c4)							R.L. n°9 - 23/02/ 1999	
Arisarum proboscideum	C (c4)					R.L. n°61 - 19/09/ 1982			
Asplenium fissum	C (c4)							R.L. n°9 - 23/02/ 1999	
Asplenium viride	C (c4)							R.L. n°9 - 23/02/ 1999	
Atropa bella- donna	C (c4)					R.L. n°61 - 19/09/ 1985	R.L. n°45 - 11/09/ 1979 and n°66 - 20/06/ 1980		
Cardamine chelidonia	C (c4)					R.L. n°61 - 19/09/ 1988			
Daphne mezereum	C (c4)						R.L. n°45 - 11/09/ 1979 and		

(continued)

Table 4 (continued)

Taxa	Classes of protection	Red List of Italian Flora (Rossi et al. 2013)	Regional Red List: Latium (Conti et al. 1997)	Regional Red List: Abruzzo (Conti et al. 1997)	Regional Red List: Molise (Conti et al. 1997)	Regional Law (Latium)	Regional Laws (Abruzzo)	Regional Law (Molise)	International Conventions
							n°66 - 20/06/ 1980		
Hyssopus officinalis subsp. aristatus	C (c4)							R.L. n°9 - 23/02/ 1999	
Ilex aquifolium	C (c4)					R.L. n°61 - 19/09/ 2001			
Nepeta nuda subsp. nuda	C (c4)				DD			R.L. n°9 - 23/02/ 1999	
Parnassia palustris subsp. palustris	C (c4)						R.L. n°45 - 11/09/ 1979 and n°66 - 20/06/ 1980		
Pulsatilla alpina subsp. millefoliata	C (c4)						R.L. n°45 - 11/09/ 1979 and n°66 - 20/06/ 1980		

Ribes multiflorum subsp. multiflorum	C (c4)					R.L. n°61 - 19/09/ 2005			
Sedum caespitosum	C (c4)					R.L. n°61 - 19/09/ 2006			
Taxus baccata	C (c4)					R.L. n°61 - 19/09/ 2009	R.L. n°45 - 11/09/ 1979 and n°66 - 20/06/ 1980		
Allium moschatum	C (c4, c5)				CR			R.L. n°9 - 23/02/ 1999	
Androsace villosa subsp. villosa	C (c4, c5)				CR			R.L. n°9 - 23/02/ 1999	
Dichoropetalum schottii	C (c4, c5)				CR			R.L. n°9 - 23/02/ 1999	
Iberis umbellata	C (c4, c5)				EN			R.L. n°9 - 23/02/ 1999	
Persicaria amphibia	C (c4, c5)				CR			R.L. n°9 - 23/02/ 1999	
Pseudofumaria alba subsp. alba	C (c4, c5)				EN			R.L. n°9 - 23/02/ 1999	

(continued)

Table 4 (continued)

Taxa	Classes of protection	Red List of Italian Flora (Rossi et al. 2013)	Regional Red List: Latium (Conti et al. 1997)	Regional Red List: Abruzzo (Conti et al. 1997)	Regional Red List: Molise (Conti et al. 1997)	Regional Law (Latium)	Regional Laws (Abruzzo)	Regional Law (Molise)	International Conventions
Quercus x pseudosuber	C (c4, c5)				CR			R.L. n°9 - 23/02/ 1999	
Ranunculus thora	C (c4, c5)				EN		R.L. n°45 - 11/09/ 1979 and n°66 - 20/06/ 1980	R.L. n°9 - 23/02/ 1999	
Veronica acinifolia	C (c4, c5)				EN			R.L. n°9 - 23/02/ 1999	
Asperugo procumbens	C (c5)		CR		EW				
Callitriche brutia	C (c5)		CR						
Linum narbonense	C (c5)		CR						
Chaerophyllum magellense	D (d1)		EW						
Helianthemum apenninum subsp. apenninum	D (d1, d4)				LR			R.L. n°9 - 23/02/ 1999	
Linum tommasinii	D (d1, d4)			LR	VU			R.L. n°9 - 23/02/ 1999	

Orchis purpurea	D (d3)								Washington Convention
Ruscus aculeatus	D (d3, d4)						R.L. n°45 - 11/09/ 1979 and n°66 - 20/06/ 1980		Habitat Directive 92/43/CEE
Asplenium ceterach subsp. bivalens	D (d4)					R.L. n°61 - 19/09/ 1983			
Cardamine graeca	D (d4)					R.L. n°61 - 19/09/ 1989			
Crepis lacera	D (d4)					R.L. n°61 - 19/09/ 1990			
Cytisus spinescens	D (d4)					R.L. n°61 - 19/09/ 1994			
Helichrysum italicum subsp. italicum	D (d4)					R.L. n°61 - 19/09/ 1999			
Juniperus deltoides	D (d4)					R.L. n°61 - 19/09/ 2002			
Turritis glabra	D (d5)		CR						
Ophrys classica	D (d3)								Washington Convention
Atriplex rosea	E (e3)			EN					

(continued)

Table 4 (continued)

Taxa	Classes of protection	Red List of Italian Flora (Rossi et al. 2013)	Regional Red List: Latium (Conti et al. 1997)	Regional Red List: Abruzzo (Conti et al. 1997)	Regional Red List: Molise (Conti et al. 1997)	Regional Law (Latium)	Regional Laws (Abruzzo)	Regional Law (Molise)	International Conventions
Barlia robertiana	E (e3)			DD	EW				Washington Convention
Blechnum spicant	E (e3)			LR	LR			R.L. n°9 - 23/02/ 1999	
Carex depressa subsp. basilaris	E (e3)			LR					
Carex ericetorum	E (e3)			LR					
Carex lasiocarpa	E (e3)								
Carpesium cernuum	E (e3)		LR						
Chrysosplenium alternifolium	E (e3)		CR	VU	EW				
Genista radiata	E (e3)			LR					
Gentiana brachyphylla subsp. favratii	E (e3)		VU	LR					
Geranium lanuginosum	E (e3)		EW						
Malus florentina	E (e3)		VU	DD	LR			R.L. n°9 - 23/02/ 1999	
Neatostema apulum	E (e3)				EW				
Orlaya daucorlaya	E (e3)			LR					
Petasites albus	E (e3)		LR						

Phyteuma hemisphaericum	E (e3)			LR	LR			R.L. n°9 - 23/02/ 1999	
Primula veris subsp. suaveolens	E (e3)								
Rorippa palustris	E (e3)			VU					
Rumex hydrolapathum	E (e3)			VU	EW				
Sanguisorba officinalis	E (e3)		EW						
Selaginella denticulata	E (e3)				EN			R.L. n°9 - 23/02/ 1999	
Valerianella discoidea	E (e3)		EW						
Vicia laeta	E (e3)		CR		CR			R.L. n°9 - 23/02/ 1999	
Alchemilla cinerea	F (f1)		LR						
Alchemilla coriacea	F (f1)		LR	LR	LR			R.L. n°9 - 23/02/1999	
Alchemilla straminea	F (f1)		LR	LR	LR			R.L. n°9 - 23/02/1999	
Alchemilla transiens	F (f1)		LR						
Alchemilla undulata	F (f1)		LR	LR					
Campanula scheuchzeri subsp. pseudostenocodon	F (f1)		EW						
Corydalis densiflora	F (f1)								

(continued)

Table 4 (continued)

Taxa	Classes of protection	Red List of Italian Flora (Rossi et al. 2013)	Regional Red List: Latium (Conti et al. 1997)	Regional Red List: Abruzzo (Conti et al. 1997)	Regional Red List: Molise (Conti et al. 1997)	Regional Law (Latium)	Regional Laws (Abruzzo)	Regional Law (Molise)	International Conventions
Euphrasia officinalis subsp. kerneri	F (f1)		EW						
Galeopsis pubescens	F (f1)		EW						
Hieracium acanthodontoides	F (f1)								
Hieracium humile subsp. brachycaule	F (f1)				LR			R.L. n°9 - 23/02/ 1999	
Hieracium latilepidotum.	F (f1)								
Hieracium marsorum	F (f1)								
Hieracium naegelianum subsp. andreae	F (f1)			LR					
Hieracium profetanum	F (f1)								
Pilosella calabra	F (f1)								
Rosa glauca	F (f1)			LR					
Saxifraga oppositifolia subsp. oppositifolia	F (f1)				LR			R.L. n°9 - 23/02/ 1999	
Saxifraga oppositifolia subsp. speciosa	F (f1)								

Seseli polyphyllum	F (f1)								
Taraxacum sect. Alpina	F (f1)				VU			R.L. n°9 - 23/02/ 1999	
Taraxacum sect. Palustria.	F (f1)		EW	EN	LR			R.L. n°9 - 23/02/ 1999	
Tolpis virgata subsp. grandiflora	F (f1)								
Alyssum cuneifolium subsp. cuneifolium	F (f2)			LR	DD				
Astragalus muelleri	F (f2)								
Bunium petraeum	F (f2)								
Cerastium thomasii	F (f2)			LR					
Dianthus cfr. guliae	F (f2)								
Leucanthemopsis alpina	F (f2)		VU	LR					
Luzula pilosa	F (f2)			LR					
Rhaponticoides centaurium	F (f2)								
Ribes petraeum	F (f2)			DD					
Ribes rubrum	F (f2)				DD				
Sambucus racemosa	F (f2)								
Saxifraga marginata	F (f2)				LR			R.L. n°9 - 23/02/ 1999	
Alcea biennis subsp. biennis	F (f3)			LR					

(continued)

Table 4 (continued)

Taxa	Classes of protection	Red List of Italian Flora (Rossi et al. 2013)	Regional Red List: Latium (Conti et al. 1997)	Regional Red List: Abruzzo (Conti et al. 1997)	Regional Red List: Molise (Conti et al. 1997)	Regional Law (Latium)	Regional Laws (Abruzzo)	Regional Law (Molise)	International Conventions
Alnus cordata	F (f3)		LR	DD					
Anethum graveolens	F (f3)		EW						
Anthriscus cerefolium	F (f3)							R.L. n°9 - 23/02/ 1999	
Pilosella corvigena	F (f3)								
Platanus orientalis	F (f3)			DD					
Rapistrum perenne	F (f3)				DD				

Table 5 Criteria for defining the protection classes of vascular flora

Classes	Criteria codes	Description of criteria
A	**a1**	*Taxa* endemic of the Park and closed areas
	a2	*Taxa* konwn in Italy only for the Park and closed areas
	a3	Rare *taxa* and/or subjected to real threats, endemic of central Apennine
	a4	Rare *taxa* or subjected to real threats exclusive, in Italy, of one of the administrative regions covered by the Park (Abruzzo, Lazio and Molise)
	a5	Rare *taxa* or subjected to real threats protected by international norms
	a6	*Taxa* included in the risk category CR (IUCN) by the National or Regional Red Lists for all administrative regions covered by the Park (Abruzzo, Lazio and Molise), or that according to recent knowledge should be included in this category
	a7	*Taxa* vey rare or under rarefaction in central Italy according to the current knowledge
B	**b1**	Not common *taxa*, endemic of one of the administrative regions covered by the Park (Abruzzo, Lazio and Molise)
	b2	Not common *taxa* exclusive, in Italy, of one of the administrative regions covered by the Park (Abruzzo, Lazio and Molise)
	b3	Rare *taxa* or of phytogeographical interest, endemic of central Apennine or known in the Park and closed areas with populations disjointed, relict or at limit of distribution area
	b4	*Taxa* included in the risk category EN (IUCN) by the National or Regional Red Lists for all administrative regions covered by the Park (Abruzzo, Lazio and Molise), or that according to recent knowledge should be included in this category
	b5	Rare *taxa* or subjected to real threats, protected by Regional laws for at least one of the administrative regions covered by the Park (Abruzzo - L.R. n°45 of 11.09.1979; n°66 of 20.06.1980; Molise - L.R. n°9 of 23.02.1999; Lazio - L.R. n°61 of 19.09.1974)
	b6	*Taxa* elsewhere common, but present in the park with a limited number of populations or individuals
C	**c1**	Italian endemics with large distribution area or common *taxa* endemic of central Apennine
	c2	*Taxa* included in the risk categories VU, LR, LC, NT (IUCN) by the National or Regional Red Lists for for at least one of the administrative regions covered by the Park (Abruzzo, Lazio and Molise) and *taxa* not common, rare or subjected to real threats, that according to recent knowledge should be included in one of this categories (excluding very common *taxa*)
	c3	Not common *taxa* and not subjected to real threats protected by international norms (excluding very common *taxa*)
	c4	Not common *taxa* and not subjected to real threats, protected by Regional laws for at least one of the administrative regions covered by the Park (Abruzzo - L.R. n°45 of 11.09.1979; n°66 of 20.06.1980; Molise - L.R. n°9 of 23.02.1999; Lazio - L.R. n°61 of 19.09.1974) (excluding very common *taxa*)

(continued)

Table 5 (continued)

Classes	Criteria codes	Description of criteria
	c5	*Taxa* included in the risk categories CR, EN (IUCN) by the National or Regional Red Lists for for at least one of the administrative regions covered by the Park (Abruzzo, Lazio and Molise) and *taxa* not common, rare or subjected to real threats, that according to recent knowledge should be included in one of this categories (excluding very common *taxa*)
D	**d1**	Very common or common *taxa* included in the risk categories VU, LR, LC, NT (IUCN) by the National or Regional Red Lists for for at least one of the administrative regions covered by the Park (Abruzzo, Lazio and Molise)
	d2	*Taxa* included in the risk categories DD (data deficent), NA (not appicable), NE (not evaluated) (IUCN) by the National or Regional Red Lists for at least one of the administrative regions covered by the Park (Abruzzo, Lazio and Molise)
	d3	Very common *taxa* and not subjected to real threats protected by international norms
	d4	Very common *taxa* and not subjected to real threats protected by Regional laws (Abruzzo - L.R. n°45 of 11.09.1979; n°66 of 20.06.1980; Molise - L.R. n°9 of 23.02.1999; Lazio - L.R. n°61 of 19.09.1974)
	d5	Very common or common *taxa* included in the risk categories CR, EN (IUCN) by the Regional Red Lists for at least one of the administrative regions covered by the Park (Abruzzo, Lazio and Molise)
E	**e1**	That have at least one of the requirements to be classified in the Classes of Protection A, B, C, D or E but to be considered Extinct (the presence of which is supported by historical herbarium specimens)
	e2	That have at least one of the requirements to be classified in the Classes of Protection A, B, C, D or E but to be considered Extinct (the presence of which is derived from historical bibliographic data)
	e3	That have at least one of the requirements to be classified in the Classes of Protection A, B, C, D or E but not confirmed in recent times
F	**f1**	That have at least one of the requirements to be classified in the Classes of Protection A, B, C, D or E, but of doubtful taxonomic status or included within critical taxonomic group of the italian vascular flora
	f2	That have at least one of the requirements to be classified in the Classes of Protection A, B, C, D or E, but the record for the Park is doubtuful
	f3	That have at least one of the requirements to be classified in the Classes of Protection A, B, C, D or E but non native plants

Melampyrum nemorosum, *Armeria arenaria* subsp. *arenaria*, *Psilurus incurvus*, *Dasiphora fruticosa*, *Saxifraga marginata*. These are mostly the result of old reports (Terracciano 1873; Tenore and Gussone 1842; Falqui 1899; Vaccari and Wilczek 1940) or a few new unlikely reports not supported by specimens (Petriglia 2004). Among the unconfirmed species, some have been found immediately outside the boundaries of the Park and are therefore likely to be confirmed soon following new research, i.e. *Phlomis herba-venti* subsp. *herba-venti*, *Buglossoides incrassata*.

From a management point of view, specific studies should be undertaken to verify the presence of the doubtful and unconfirmed plants which make up 7.4 % of the flora in the Park.

Non-native Species

The non-native plants amount to 90 *taxa*, 4.2 % of the flora in the Park. Among these, 36 are casual, 7 invasive (*Ailanthus altissima*, *Fallopia baldschuanica*, *Isatis tinctoria* subsp. *tinctoria*, *Robinia pseudoacacia*, *Senecio inaequidens*, *Veronica persica*, *Vitis* x *koberi*) and 47 naturalized. Invasive plants must be monitored with the aim of starting the containment or complete eradication of *Senecio inaequidens*, *Ailanthus altissima* and *Robinia pseudoacacia*.

Acknowledgements We would like to thank the Director and the employers of National Park which have sponsored and supported this work, to F. Pedrotti for having drawn up the chapter on vegetation.

We thank the colleagues who kindly assisted us with the identification of critical specimens: E. Lattanzi (*Rosa*), D. Marchetti (*Pteridophyta*), G. Domina (*Orobanche*), G. Tondi (*Alchemilla*), G. Gottschlich (*Hieracium*, *Pilosella*), K. Pagitz (*Rubus*), C. Zidorn (*Leontodon*)), E. Banfi (*Bromopsis*).

We gratefully acknowledge the Directors and Curators of FI, RO, NAP.

Special thanks go to the friends S. Gregg for the linguistic revision and G. Serafini for providing us some digital images, to N. Ranalli and R. P. Wagensommer for helping us in the compilation of the floristic data base.

References

Aedo D, Garcia MA, Alarcon ML, Aldasoro JJ, Navarro C (2007) Taxonomic revision of Geranium subsect. Mediterranea (Geraniaceae). Syst Bot 32(1):93–128

Aleffi M (1992) Associazioni di Briofite ed Alghe dei laghi artificiali di Campotosto e Barrea (Abruzzo, Italia centrale). Doc Phytosoc 14:91–96

Anzalone B (1961) Su alcune piante interessanti di Scanno e di altre località d'Abruzzo. Nuovo Giorn Bot Ital 67(3–4):550–556

Anzalone B (1962) Su alcune piante nuove o interessanti per il Lazio, l'Abruzzo o altre regioni dell'Italia Centrale. Ann Bot (Roma) 27(2):339–359

Anzalone B, Bazzichelli G (1960) La flora del Parco Nazionale d'Abruzzo. Ann Bot (Roma) 26(2–3):1–182 (reprint)

Anzalone B, Veri L (1975) Su alcune piante interessanti o nuove per Lazio e Abruzzo. Giorn Bot Ital 109(4–5):251–255

Ardenghi NMG, Galasso G, Banfi E, Zoccola A, Foggi B, Lastrucci L (2014) A taxonomic survey of the genus Vitis L. (Vitaceae) in Italy, with special reference to Elba Island (Tuscan Archipelago). Phytotaxa 166(3):163–198

Avena GC, Rosati M (1974) Sulla presenza di Geum rivale L. in località di Pantano di Scanno; primo rinvenimento per la regione Abruzzo. Ann Bot (Roma) 32:97–112

Banfi E, Galasso G (2014) Notulae alla checklist della flora vascolare italiana 17 2063. Inform Bot Ital 46(1):81

Ballelli S, Lucarini D, Pedrotti F (2005) Catalogo dell'Erbario dei Monti Sibillini di Vittorio Marchesoni. Braun-Blanquetia 38:1–257

Bartolucci F (2010) Verso una revisione biosistematatica del genere Thymus L. in Italia: considerazioni nomenclaturali, sistematiche e criticità tassonomica. Ann Bot (Roma) (2009):135–148

Bartolucci F, Peruzzi L (2007) Distribuzione del genere Gagea Salisb. (Liliaceae) nell'Appennino centro-settentrionale. Biogeographia 28:205–238

Bartolucci F, Peruzzi L (2014) Thymus paronychioides Čelak. (Lamiaceae), a neglected species from Sicily belonging to section Hypodromi. Folia Geobot 49(1):83–106

Bartolucci F, Conti F, Iamonico D, Del Guacchio E (2014a) A new combination in Mcneillia (Caryophyllaceae) for the Italian flora. Phytotaxa 170(2):139–140

Bartolucci F, Peruzzi L, Soldano A (2014b) Notulae alla checklist della flora vascolare italiana 17, 2064–2069. Inform Bot Ital 46(1):81–83

Bassani P (1994) Contributo alla flora del Parco Nazionale d'Abruzzo. Ann Bot (Roma) 50(suppl 9):107–110

Baumann B, Baumann H (1988) Ein Beitrag zur Kenntnis der Gattung Epipactis Zinn in Mittelmeergebiet, 20th edn. Mitt Arebeitskreis Heimische Orchid Baden, Württemberg, pp 1–68

F. Conti, F. Bartolucci, *The Vascular Flora of the National Park of Abruzzo, Lazio and Molise (Central Italy)*, Geobotany Studies, DOI 10.1007/978-3-319-09701-5

Baumann H, Lorenz R (1988) Beiträge zur Kenntnis der Gattung Epipactis Zinn in Mittel-und Süditalien und der Verbreitung einiger in diesem Gebiet spät blühenden Orchideen, vol 3, 20th edn. Mitt Arbeitskreis Heimische Orchid, Baden-Württemberg, pp 652–694

Bazzichelli G (1972) Achillea barrelieri (Ten.) Sch.-Bip. (emend.Heimerl) ssp. barrelieri forma schouwii (DC.) Bazzichelli n. comb. (Compositae): Sviluppo del gametofito femminile. Revisione sistematica e nomenclatura. Distribuzione geografica. Ann Bot (Roma) 29:31–85

Bazzichelli G, Furnari F (1970) Ricerche sulla flora e sulla vegetazione di altitudine nel Parco Nazionale d'Abruzzo. Pubbl Ist Bot Univ Catania, pp 1–41

Bertoloni A (1833–1854) Flora Italica 1–10. Tip. R. Masi, Bologna

Betti S (1997) Una settimana in Abruzzo. GIROS Notizie 6:7–10

Biondi E, Allegrezza M, Frattaroli R (1992) Inquadramento fitosociologico di alcune formazioni pascolive dell'Appennino Abruzzese-Molisano. Doc Phytosoc 14:195–210

Bohn U, Gollub G, Hettwer C (2000) Karte der natürlichen Vegetation Europas. Maßstab 1:2.500.000. Bundesamt für Naturschutz, Bonn-Bad Godesberg

Boni C, Bono P, Capelli G (1986) Schema idrogeologico dell'Italia Centrale. Mem Soc Geol It 35:991–1012

Bortolotti L (1965) Il Parco Nazionale d'Abruzzo. In: I Parchi Nazionali d'Italia. Ist di Tecnica e Propaganda Agraria, Roma

Brullo C, Brullo S, Giusso del Galdo G (2010) Note tassonomiche e nomenclaturali su Allium permixtum (Alliaceae), specie critica della flora italiana. In: Atti del convegno: La biodiversità vegetale in Italia: aggiornamenti sui gruppi critici della flora vascolare. Dip Biol Amb, La Sapienza Università di Roma, 22–23 Ottobre 2010

Bruno F, Bazzichelli G (1966) Note illustrative alla carta della vegetazione del Parco Nazionale d'Abruzzo (scala 1: 25000). Progetto conservazione geobotanico. Ann Bot (Roma) 28(3):739–778

Bruno F, Bazzichelli G (1968) Carta della vegetazione del Parco Nazionale d'Abruzzo. Ente autonomo Parco Nazionale d'Abruzzo, Roma

Buchwald R (1995) Vegetazione e odonatofauna negli ambienti acquatici dell'Italia centrale. Braun-Blanquetia 11:1–77

Canullo R, Pedrotti F (1992) Processi e tendenze dinamiche nella vegetazione delle Mainarde (Val Pagana-Le Forme). L'Uomo e l'Ambiente 16:121–123

Canullo R, Pedrotti F (1993) The cartographic representation of the dynamic tendencies in the vegetation: a case study from the Abruzzo National Park (Italy). Oecologia Montana 2:13–18

Celesti-Grapow L, Pretto F, Carli E, Blasi C (eds) (2010) Flora vascolare alloctona e invasiva delle regioni d'Italia. Centro Stampa Università "La Sapienza", Roma

Christenhusz MJM, Govaerts R, David JC, Hall T, Borland K, Roberts PS, Tuomisto A, Buerki S, Chase MW, Fay MF (2013) Tiptoe through the tulips – cultural history, molecular phylogenetics and classification of Tulipa (Liliaceae). Bot J Linn Soc 172(3):280–328

Christenhusz MJM, Zhang X-C, Schneider H (2011a) A linear sequence of extant families and genera of lycophytes and ferns. Phytotaxa 19:7–54

Christenhusz MJM, Reveal JL, Farjon A, Gardner MF, Mill RR, Chase MW (2011b) A new classification and linear sequence of extant gymnosperms. Phytotaxa 19:55–70

Conti F (1992) Alcune piante di particolare interesse fitogeografico rinvenute sulle Mainarde (Lazio e Molise). In: Pedrotti F, Tassi F (eds) Le Mainarde. Zona d'ampliamento in Molise del Parco Nazionale d'Abruzzo. L'uomo e l'Ambiente 16:81–97

Conti F (1994) Su alcune piante nuove o notevoli per la Flora del Parco Nazionale d'Abruzzo. Ann Bot (Roma) 50(9):97–105

Conti F (1995) Prodromo della Flora del Parco Nazionale d'Abruzzo. In: Tassi F (ed) Progetto Biodiversità. Ente Autonomo del Parco Nazionale d'Abruzzo. Almadue srl, Roma

Conti F (1998) An annotated checklist of the flora of the Abruzzo. Bocconea 10

Conti F (2004) Suddivisioni fitogeografiche della regione Abruzzo. Coll Phytosoc 28:731–745 (Berlin-Stuttgart)

Conti F, Abbate G, Alessandrini A, Blasi C (2005) An annotated checklist of the Italian vascular Flora. Palombi Editore, Roma

Conti F, Alessandrini A, Bacchetta G, Banfi E, Barberis G, Bartolucci F, Bernardo L, Bonacquisti S, Bouvet D, Bovio M, Brusa AG, Del Guacchio E, Foggi B, Fratini S, Galasso G, Gallo L, Gangale C, Gottschlich G, Grünanger P, Gubellini L, Iriti G, Lucarini D, Marchetti D, Moraldo B, Peruzzi L, Poldini L, Prosser F, Raffaelli M, Santangelo A, Scassellati E, Scortegagna S, Selvi F, Soldano A, Tinti D, Ubaldi D, Uzunov D, Vidali M (2007) Integrazioni alla checklist della flora vascolare italiana. Natura Vicentina 10:5–74

Conti F, Bartolucci F (2011a) Notulae alla checklist della flora vascolare italiana 11. 1782–1793. Inform Bot Ital 43(1):132–135

Conti F, Bartolucci F (2011b) Notulae alla checklist della flora vascolare italiana 12. 1858–1864. Inform Bot Ital 43(2):365–366

Conti F, Bartolucci F, Catonica C, D'Orazio G, Londrillo I, Manzi A, Tinti D (2006) Aggiunte alla flora d'Abruzzo. II° contributo. Inform Bot Ital 38(1):113–116

Conti F, Bartolucci F, Iocchi M, Tinti D (2011a) Atlas of the pteridological knowledge of Abruzzo (Central Italy). Webbia 66(2):251–305

Conti F, Bartolucci F, Manzi A, Miglio M, Tinti D (2008) Aggiunte alla Flora d'Abruzzo: III contributo. Ann Mus Civ Rovereto Sez:Arch St Sc Nat 23:127–140

Conti F, Di Carlo F, Manzi A, Paolucci M (2011b) Notulae alla checklist della flora vascolare italiana 11. 1799–1802. Inform Bot Ital 43(1):136–137

Conti F, Guarrera P, Manzi A, Pellegrini M (1987) Nuove stazioni di Juniperus sabina L. per la Majella e Parco Nazionale d'Abruzzo, sua distribuzione nell'Italia Centrale e impieghi tradizionali. Inform Bot Ital 18(1-2-3):117–122

Conti F, Iamonico D (2013) Notulae alla checklist della flora vascolare italiana 16. 2009. Inform Bot Ital 45(2):303

Conti F, Manzi A (1992) A new association of the calcareous screes in the Mainarde mountains. Doc Phytosoc 14:499–504

Conti F, Manzi A, Pedrotti F (1992) Libro Rosso delle Piante d'Italia. WWF Italia. TIPAR Poligrafica Editrice, Roma

Conti F, Manzi A, Pedrotti F (1997) Liste Rosse Regionali delle Piante d'Italia. WWF Italia. Società Botanica Italiana. Università di Camerino. Camerino

Conti F, Manzi A, Pirone G (1999) Note floristiche per l'Abruzzo. Inform Bot Ital 30(1–3):15–22 (1998)

Conti F, Manzi A, Tinti D (2002) Aggiunte alla Flora d'Abruzzo. Inform Bot Ital 34(1):55–61

Conti F, Miglio M, Santucci B (2011c) Notulae alla checklist della flora vascolare italiana 11. 1797–1798. Inform Bot Ital 43(1):136

Conti F, Minutillo F (1998) Aggiunte e rettifiche alla Flora del Parco Nazionale d'Abruzzo. Ann Bot (Roma) 54(2):97–113

Conti F, Minutillo F (2001) Nuove aggiunte alla flora del Parco Nazionale d'Abruzzo. Inform Bot Ital 33(1):11–13

Conti F, Pedrotti F, Pirone G (1990) Su alcune piante notevoli rinvenute in Abruzzo, Molise e Basilicata. Arch Bot Ital 66(3–4):182–196

Conti F, Pellegrini M (1990) Orchidee spontanee d'Abruzzo. Cogecstre Ed, Lanciano

Conti F, Peruzzi L (2006) Pinguicula (Lentibulariaceae) in Central Italy: taxonomic study. Ann Bot Fennici 43:321–337

Conti F, Soldati R (2010) Notule Pteridologiche Italiche. VIII: 188. Dryopteris submontana (Fraser-Jenk. & Jermy) Fraser-Jenk. Ann Mus Civ Rovereto Sez: Arch St Sc Nat 25:108

Conti F, Tinti D (2012) Flora vascolare della Riserva Naturale "Gole del Sagittario" (Abruzzo). Boll Mus Civico Storia Nat Verona 36:3–30

Conti F, Tinti D, Bartolucci F, Scassellati E, Di Santo D, Fanelli C, Iocchi M, Meister J, Pavoni P, Torcoletti S (2010) Banca Dati della Flora Vascolare d'Abruzzo: lo stato dell'arte. Ann Bot (Roma) suppl 2009:85–94

Corazzi G, Lattanzi E, Tilia A (2003) Note su Orobanche ebuli Huter et Rigo. Inform Bot Ital 35 (1):3–6

Corbetta F, Pirone G (1989) La vegetazione del fiume Tirino (Abruzzo). Arch Bot Ital 65 (3–4):121–153

Crugnola A (1894) La vegetazione al Gran Sasso d'Italia. G. Fabbri, Teramo

D'Andrea M (1982) Le piante officinali del Parco Nazionale d'Abruzzo e gli usi popolari di esse nell'alta valle del Sangro. Rivista Abruzzese 35:155–176

D'Andrea M, Pantaloni M, Praturlon A (2003) Itinerario N°14. Da Sora a Castel San Vicenzo. In: Società Geologica I (ed) Guide Geologiche Regionali, Abruzzo. BE.MA editrice, Milano, pp 268–301

Del Prete C, Donini AM, Garbari F (1981) Quisquiliae Floristicae Apenninae: 1–5. Atti Soc Tosc Sci Nat Pisa Mem Ser B 87:71–84

Dillenberger MS, Kadereit JW (2014) Maximum polyphyly: multiple origins and delimitation with plesiomorphic characters require a new circumscription of Minuartia (Caryophyllaceae). Taxon 63:64–88

Di Pietro R, De Santis A, Fortini P (2005) A geobotanical survey on acidophilous grasslands in the Abruzzo, Lazio and Molise National Park (Central Italy). Lazaroa 26:115–137

Di Pietro R, Proietti S, Fortini P, Blasi C (2004) La vegetazione dei ghiaioni del settore Sud-orientale del Parco Nazionale d'Abruzzo, Lazio e Molise. Fitosociologia 41(2):3–20

Falqui G (1899) Contributo alla Flora del bacino del Liri. Atti Accad Scienze fis e nat di Napoli, ser 2, 9:1–51

Ferrari C, Rossi G (1985) Preliminary observations on the summer diet of the Abruzzo chamois (R. rupicapra ornata Neumann 1899). In: Lovari S (ed) Biology and management of mountain ungulate. Croom Helm, London, pp 85–92

Ferrari C, Rossi G, Cavani C (1988) Summer food habits and quality of female, kid and subadult Apennine chamois, Rupicapra pyrenaica ornata Neumann 1899 (Arctiodactyla, Bovidae). Z Säugetierkd 53:170–177

Ferrarini E, Ciampolini F, Pichi Sermolli REG, Marchetti D (1986) Iconographia Palynologica Pteridophytorum Italiae. Webbia 40(1):1–202

Fiori A (1927) Escursioni botaniche in Abruzzo (Resoconto dell'adunanza del giorno 14 maggio 1927). Nuovo Giorn Bot Ital 34:778–779

Fiori A (1928) Nuova Flora Analitica d'Italia 2(6):927. Edagricole, Firenze

Fiori A (1943) Flora Italica Criptogama. Pars V: Pteridophyta. Tipografia Mariano Ricci, Firenze

Fiori A, Béguinot A (1927) Schedae ad Floram Italicam Exsiccatam. Series III. Centuriae XXIX–XXX, 16th edn. Tip. Valbonesi, Messina, pp 337–435

Fiori A, Béguinot A, Pampanini R (1907) Schedae ad Floram Italicam Exsiccatam. Centuriae VI–VII. Nuovo Giorn Bot Ital 14(2):247–291

Furrer E (1928) Die Höhenstufen des Zentralapennin. Vierteljiahrsschrift Naturfosch Ges Zürich 73:642–663

Garbari F, Peruzzi L, Tornadore N (2007) Ornithogalum L. (Hyacinthaceae Batsch) e generi correlati (subfam. Ornithogaloideae Speta) in Italia. Atti Soc Tosc Sci Nat Mem B 114:35–44

Giancola C, Stanisci A (2006) La vegetazione delle rupi di altitudine del Molise. Fitosociologia 43 (1):187–195

Gottschlich G (2009) Die Gattung Hieracium (Compositae) in der region Abruzzen (Italien). Stapfia 89:1–328

Grande L (1904) Primo contributo alla Flora di Villavallelonga nella Marsica. Nuovo Giorn Bot Ital 11(2):125–140

Grande L (1910) Note di Floristica Napoletana. Bull Orto Bot Napoli 2:513–520

Grande L (1912) Note di Floristica Napoletana. Bull Soc Bot Ital 175–186

Grande L (1913) Note di Floristica Napoletana. VIII–XL. Bull Orto Bot Napoli 3:193–218

Grande L (1914) Note di Floristica. Bull Orto Bot Napoli 4:363–370

Grande L (1916) Note di Floristica. Bull Orto Bot Napoli 5:55–67

Grande L (1924) Note di Floristica. Boll Soc Nat Napoli 36:217–245

Grande L (1925) Note di Floristica. Nuovo Giorn Bot Ital 32:62–101

Gravina P (1812) Giornale della peregrinazione Botanica eseguita nelle Montagne del Circondario di Scanno, dal Sig. Pasquale Gravina. Giornale Enciclopedico di Napoli 6:3–49

Greco S, Petriccione B (1989) La cartografia della vegetazione nella definizione della qualità dell'ambiente: il caso di Cocullo (AQ). Not Fitosoc 24:63–98

Griebl N (2010) Die Orchideen der Abruzzen. Ber Arbeitskrs Heim Orchid 27(2):112–159

Guarrera P, Tammaro F (1991) Aspetti naturalistici dei Monti della Laga e di altri territori montani circostanti. In: La Valle dell'Alto Vomano e i Monti della Laga, vol 1. CARSA Ed, Pescara, pp 40–63

Hamasha HR, Von Hagen KB, Röser M (2012) Stipa (Poaceae) and allies in the old world: molecular phylogenetics realigns genus circumscription and gives evidence on the origin of American and Australian lineages. Plant Syst Evol 298(2):351–367

Hempel W (2011) Revision und Phylogenie der Arten der Gattung Melica L. (Poaceae) in Eurasien und Nordafrika. Feddes Repert 122(1–2):1–253

Hennecke G, Hennecke M (1999) Neue Orchideen-Funde in den Abruzzen. J Eur Orchid 31 (4):936–948

Hoffmann V (1989) Orchideenfunde in Marken und Abruzzen (Italien) in der Zeit vom 12.-16.8.1986. Ber Arbeitskreis Heimische Orchid 6:101–105

Iamonico D (2009) Aggiornamenti foristici per il genere Amaranthus L. (Amaranthaceae) in Italia. Inform Bot Ital 41(2):303–306

Iamonico D, Bartolucci F, Conti F (2011) Notulae alla Flora Esotica d'Italia 5. 93. Inform Bot Ital 43(2):373

Koopman J, Smith C, Blackstock N (2014) Carex tomentosa versus C. filiformis (Cyperaceae): the long-standing debate comes to its happy end. Nordic J Bot 32(5):667–670

Lastoria M (2000) Flora d'Abruzzo, 2. Deltagrafica, Teramo

Lattanzi E, Minutillo F, Tilia A (2000) Segnalazioni Floristiche Italiane: 932. Orobanche ebuli Huter & Rigo (Orobanchaceae). Inform Bot Ital 31(1–3):81

Loche P, Leone M, Squartini V (1992) Segnalazioni Floristiche Italiane: 625–626. Inform Bot Ital 23(1):50–51

Lorito FMG, Veri L (1975) Ad floram italicam notulae taxonomicae et geobotanicae. 15. La Falcaria vulgaris Bernh. in Abruzzo. Webbia 29(2):539–544

Lusina G (1954) Il Parco Nazionale d'Abruzzo. Nuovo Giorn Bot Ital 60(4):870–873

Manzi A (1990) La gestione dei pascoli montani in Abruzzo e la Società delle Erbe Seconde di Pescasseroli ed Opi. Arch Bot Ital 66(3–4):129–142

Manzi A (1993) Note floristiche per le regioni Abruzzo e Marche. Arch Bot Ital 68(3–4):173–180

Manzi A, Conti F (2002) Le sorgenti del Vomano: un ambiente umido di grande interesse floristico. In: Clementi A, Osella B (eds) Chiarino, rocce, piante, animali, uomini. Le Orme, L'Aquila

Marchetti D (2004) Le Pteridofite d'Italia. Ann Mus Civ Rovereto Sez: Arch St Sc Nat 19:71–231

Marchi P (1972) Per una revisione citotassonomica della Flora Italiana. Genere Leucanthemum Adams. em. Briq. & Cav. (Compositae): "Osservazioni su specie appartenenti al ciclo di L. vulgare Lam.". Ann Bot (Roma) 29:259–295

Marchi P, Illuminati O (1974) Notizie e considerazioni sui Leucanthemum (Compositae) della flora d'Italia. Ann Bot (Roma) 33:167–194

Mennema J (1989) A taxonomic revision of Lamium (Lamiaceae). Leiden Bot Ser 11:1–198

Minutillo F (1995) Segnalazioni Floristiche Italiane: 773. Inform Bot Ital 26(2–3):223

Montelucci G (1958) Appunti sulla vegetazione del Monte Velino (Appennino Abruzzese). Nuovo Giorn Bot Ital 65(1–2):237–334

Montelucci G (1962) Itinerario geobotanico da Tivoli all'Aquila. Nuovo Giorn Bot Ital 68 (3–4):335–375

Murbeck S (1933) Monographie der Gattung Verbascum. Lunds Univ Arsskrift 20:1–630

Naviglio L (1984) Aspetti naturalistici del Lago Pantaniello. Natura e Montagna 31(3):49–57

Orsomando E (1975) La distribuzione dell'Epipogium aphyllum nell'Appennino con due nuove stazioni nel Parco Nazionale d'Abruzzo e nei Monti della Laga. Arch Bot Biogeogr Ital 19:171–180

Paura B, Abbate G (1995) I querceti caducifoglie del Molise: primo contributo sulla sintassonomia e corologia. Ann Bot (Roma) 51(10(2)):325–339

Pedrotti F (1983) Alcuni ambienti umidi del Molise. Giorn Bot Ital 117(1):131–132

Pedrotti F (1991) Carta della vegetazione reale d'Italia. In: Relazione sullo stato dell'ambiente, Ministero dell'Ambiente, Roma

Pedrotti F (1996) Il pioppo tremulo (*Populus tremula* L.) nella colonizzazione dei terreni abbandonati del parco nazionale d'Abruzzo. Coll Phytosoc 24:111–122

Pedrotti F, Gafta D (1996) Ecologia delle foreste ripariali e paludose dell'Italia. L'Uomo e l'Ambiente 23:1–165

Pedrotti F, Gafta D, Manzi A, Canullo R (1992) Le associazioni vegetali della Piana di Pescasseroli (Parco Nazionale d'Abruzzo). Doc Phytosoc 14:123–147

Pedrotti F, Spada F, Conti F (1996) Tipificazione di una nuova associazione a Salix apennina dell'Appennino centrale. L'Uomo e l'Ambiente 23:153

Peruzzi L, Bartolucci F (2006) Gagea luberonensis J.-M. Tison (Liliaceae) new for the Italian flora. Webbia 61(1):1–12

Peruzzi L, Bartolucci F, Conti F (2014a) Notulae alla checklist della flora vascolare italiana 18. 2099–2100. Inform Bot Ital 46(2) (in press)

Peruzzi L, Conti F, Bartolucci F (2014b) An inventory of vascular plants endemic to Italy. Phytotaxa 168:1–75

Petriccione B (1986) Una nuova stazione di Leontopodium nivale (Ten.) Huet sull'Appennino Centrale. Ann Bot (Roma) 43:151–156

Petriccione B (1988) Segnalazioni floristiche per il Parco Nazionale d'Abruzzo. Ann Bot (Roma) 44(4):159–166

Petriccione B, Greco S, Tammaro F (1994) La vegetazione del progettato Parco Archeologico-Naturalistico della Valle di Amplero e della Vallelonga (AQ). Micol Veg Medit 8(2):137–160

Petriglia B (2004) Flora illustrata della Ciociaria. Provincia di Frosinone, Assessorato alla Pianificazione territoriale. CD-ROM

Pignatti S (1976) Tavoletta Pescocostanzo. Foreste, pascoli e coltivi. Associazioni vegetali. Vegetazione potenziale, vol 3. Carta della Montagna, Roma, Ministero Agricoltura Foreste, Roma, pp 444–457, 480–482

Pignatti S (1982) Flora d'Italia, 1-3. Edagricole, Bologna

Pignotti L (2003) Scirpus L. and related genera (Cyperaceae) in Italy. Webbia 58(2):281–400

Pirone G (1997) Il paesaggio vegetale di Rivisondoli: aspetti della flora e della vegetazione. Edigrafital, Teramo

Pirone G, Corbetta F, Frattaroli AR, Tammaro F (1997) Studi sulla Valle Peligna (Italia centrale, Abruzzo): la copertura vegetale. Quaderni di Provincia Oggi 23(1):81–119

Pirone G, Frattaroli AR, Ciaschetti G (2010) Le serie di vegetazione della Regione Abruzzo. In: Blasi C (ed) La vegetazione d'Italia con carta delle serie di vegetazione in scala 1: 500.000. Palombi Editori, Roma, pp 311–335

Pirone G, De Nuntiis P (2002) A new plant association of the calcareous moist rocks of the Apennines in the Abruzzo region (Italy). Plant Biosyst 136(1):83–90

Pirone G, Tammaro F (1997) The hilly calciophilous garigues in Abruzzo (Central Apennines-Italy). Fitosociologia 32:73–90

Pirone G, Tammaro F (1998) La biodiversità vegetale in Abruzzo e il suo stato di conservazione. In: Burri E (ed) Aree protette in Abruzzo. Contributi alla conoscenza naturalistica ed ambientale: 77–119. Università dell'Aquila-Dip. Scienze Ambientali-Reg. Abruzzo. Carsa, Pescara

Pirotta R (1933) La flora. In: Il Parco Nazionale d'Abruzzo. Club Alpino Italiano, Roma, pp 79–83

Praturlon A, Miccadei E, Piacentini T (2003) Itinerario N°10. Da Pescina ad Anversa degli Abruzzi. In: Società Geologica I (ed) Guide Geologiche Regionali, Abruzzo. BE.MA editrice, Milano, pp 212–245

Reinhard HR (1987) Untersuchungen an Ophrys holoserica (Burm. fil.) W. Greuter subsp. elatior (Gumprecht) Gumprecht (Orchidaceae). Mitt Arbeitskreis Heimische Orchid Baden-Württemberg 19:769–800

Reveal JL, Chase MW (2011) APG III: bibliographical information and synonymy of Magnoliidae. Phytotaxa 19:71–134

Rossi W, Bassani P (1982) Una nuona stazione di Nigritella nigra dell'appennino centrale. Atti Soc Tosc Sci Nat Pisa Mem B 87:225–228

Rossi W, Klein E (1987) Eine neue Unterart der Epipactis helleborine (L.) Crantz aus Mittelitalien: Epipatis helleborine (L.) Crantz ssp. latina W. Rossi & E. Klein subspecies nova. Orchidee (Hamburg) 38(2):93–95

Rossi G, Montagnani C, Gargano D, Peruzzi L, Abeli T, Ravera S, Cogoni A, Fenu G, Magrini S, Gennai M, Foggi B, Wagensommer RP (2013) Lista Rossa della Flora Italiana. 1. Policy Species e altre specie minacciate. Comitato Italiano IUCN e Ministero dell'Ambiente e della Tutela del Territorio e del Mare

Rovelli E (1992) Segnalazioni Floristiche Italiane: 635. Inform Bot Ital 23(1):55

Rovelli E, Conti F (1995) Note floristiche per l'Appennino Centrale. Arch Geobot 1(2):185–188

Rovesti G, Rovesti P (1934) Flora officinale del Parco Nazionale d'Abruzzo e delle zone limitrofe. Riv Ital Essenze Profumi ecc 16:197–221

Schlee M, Göker M, Grimm GW, Hemlben V (2011) Genetic patterns in the Lathyrus pannonicus complex (Fabaceae) reflect ecological differentiation rather than biogeography and traditional subspecific division. Bot J Linn Soc 165:402–421

Scoppola A, Modena M (1997) Aspetti fitosociologici delle faggete di Collelongo (AQ). Italia Italia Forestale Montana 52(2):102–117

Sipari E (1926) Relazione del presidente del direttorio provvisorio dell'Ente Autonomo del Parco Nazionale d'Abruzzo alla commissione amministratrice dell'ente stesso, nominata con Regio Decreto 25 marzo 1923. Tip. Majella di A, Tivoli

Slovák M, Kučera J, Marhold K, Zozomová-Lihová J (2012) The morphological and genetic variation in the polymorphic species Picris hieracioides (Compositae, Lactuceae) in Europe strongly contrasts with traditional taxonomical concepts. Syst Bot 37:258–278

Società Botanica Italiana (1953) Escursione sociale al M. Terminillo e al Parco Nazionale d'Abruzzo (30 luglio-2 agosto 1953). Nuovo Giorn Bot Ital 60:858–885

Soster M (2001) Identikit delle Felci d'Italia. Guida al riconoscimento delle Pteridofite italiane. Valsesia Editrice, Italy

Spada F (1979) Nuove segnalazioni per la flora del Parco Nazionale d'Abruzzo. Arch Bot Biogeogr Ital 54(3–4):154–162

Spada F, Conti F (1994) Lagozzo: patterns of floristic diversisity and topographical heterogeneity in a forest ecosistem in Monti della Meta (S-C Italy). Giorn Bot Ital 128(1):385

Stanisci A (1994) High-mountain dwarf shrublands in Abruzzo National Park and Majella massif: preliminary results. Fitosociologia 26:81–91

Stanisci A (1997) Gli arbusteti altomontani dell'Appennino Centrale e Meridionale. Fitosociologia 34:3–46

Steffan M, Steffan P (1986) Segnalazioni Floristiche Italiane: 311–313. Inform Bot Ital 17(1-2-3):120–121

Steinberg C (1971) Revisione sistematica e distributiva delle Adonis annuali in Italia. Webbia 25 (2):299–351

Tammaro F (1984a) Carex nuove o rare per la Flora d'Abruzzo o dell'Italia Centro-Meridionale. Arch Bot Biogeogr Ital 59(3–4):175–178

Tammaro F (1984b) Segnalazioni Floristiche Italiane: 247–254. Inform Bot Ital 15(1):86–89

Tammaro F (1986) Segnalazioni Floristiche Italiane: 275–278. Inform Bot Ital 16(2–3):261–272

Tammaro F (1988) La distribuzione del genere Carex L. (Cyperaceae) in Abruzzo. Inform Bot Ital 19(3):287–304

Tammaro F (1998) Il paesaggio vegetale d'Abruzzo. Aree protette, biotopi ed itinerari botanici: dalle zone costiere ai massicci montuosi. Cogecstre Edizioni, Penne

Tammaro F, Pace L (1994) Considerazioni floristiche sulla Conca del Fucino. In: VV.AA., Il lago Fucino e il suo emissario. Carsa Edizioni, Pescara:78–95

Tammaro F, Sabatini L, Mastracci M (1988) Cartografia floristica di entità della Flora d'Abruzzo: le Ombrellifere. Boll Ass Ital Cart 72-73-74:709–725

Tammaro F, Visca C (1987) Segnalazioni Floristiche Italiane: 465–477. Inform Bot Ital 19 (2):181–184

Tenore M (1830) Succinta relazione del viaggio fatto in Abruzzo ed in alcune parti dello Stato Pontificio dal Cavalier Tenore nell'Està del 1829. Stamperia della Società Filomatica:[1]-90 [91]

Tenore M (1831) Sylloge plantarum vascularium Florae Neapolitanae hucusque detectarum. Ex Typ. Fibreni, Neapoli

Tenore M (1835) Ad Florae Neapolitanae Syllogem, Appendix quarta.Tipografia del Fibreno, Neapoli

Tenore M (1835–1838) Flora Napolitana 5. Stamperia e Cartiera del Fibreno, Napoli

Tenore M (1842) Ad Florae Neapolitanae Syllogem, Appendix quinta. Typis p. Tizzano, Neapoli

Tenore M, Gussone G (1842) Memorie sulle peregrinazioni eseguite dai soci ordinari Signori M. Tenore e G. Gussone. Stamperia Reale, Napoli

Terracciano N (1872) Relazione intorno alle peregrinazioni botaniche fatte nella provincia di Terra di Lavoro. Nobili & Cie, Caserta

Terracciano N (1873) Seconda relazione intorno alle peregrinazioni botaniche fatte nella provincia di Terra di Lavoro. Nobile e C., Caserta

Terracciano N (1874) Terza relazione intorno alle peregrinazioni botaniche fatte nella provincia di Terra di Lavoro. Nobile e C., Caserta

Terracciano N (1878) Quarta relazione intorno alle peregrinazioni botaniche fatte nella provincia di Terra di Lavoro. Nobile e C., Caserta

Terracciano N (1890) Intorno ad alcune piante della flora di Terra di Lavoro. Napoli

Tison J-M, Peterson A, Harpke D, Peruzzi L (2012) Reticulate evolution of the critical Mediterranean Gagea sect. Didymobulbos (Liliaceae), and its taxonomic mplications. Plant Syst Evol 299(2):413–438. doi:10.1007/s00606-012-0731-4

Vaccari L (1911) Plantae Italicae Criticae. Ann Bot (Roma) 9:15–37

Vaccari L, Wilczek E (1940) Contributo alla conoscenza floristica del Parco Nazionale d'Abruzzo. Chanousia 4:179–198

Venanzoni R (1987) Segnalazioni Floristiche Italiane: 503. Inform Bot Ital 19(2):195

Vezzani L, Ghisetti F (1998) Carta geologica dell'Abruzzo. scale 1:100.000. S.EL.CA., Firenze

Viegi L, Cela Renzoni G, D'Eugenio ML, Rizzo AM (1990) Flora esotica d'Italia: le specie presenti in Abruzzo e in Molise (revisione bibliografica e d'erbario). Arch Bot Ital 66(1–2):1–128

Zanotti AL, Cristofolini G (1994) Taxonomy and chorology of Helleborus L. sect. Helleborastrum Spach in Italy. Webbia 49(1):1–24

Zodda G (1931) Prime notizie sulla Flora delle Mainarde. Ann Bot (Roma) 19:163–201

Printed by Printforce, the Netherlands